Advances in Porcelain Enamel Technology

Advances in Porcelain Enamel Technology

Ceramic Transactions, Volume 211

A Collection of Papers Presented at the 70th and 71st Porcelain Enamel Institute Technical Forum, June 9–12, 2008, May 4–7, 2009, Nashville, TN

Conference Director
Peter Vodak

Assistant Conference Director
Mike Horton

Chief Editor
Charles Baldwin

Assistant Editors
2008—Holger Evele
2009—Renee Pershinsky

The American Ceramic Society

WILEY

A John Wiley & Sons, Inc., Publication

Published by John Wiley & Sons, Inc., Hoboken, New Jersey.
Published simultaneously in Canada.

Library of Congress Cataloging-in-Publication Data is available.

ISBN 978-0-470-40841-4

Printed in the United States of America.

10 9 8 7 6 5 4 3 2 1

Contents

Preface

The PEI Technical Forum Committee is pleased to provide you with the combined proceedings of the 70th and 71st Annual PEI Technical Forum. This volume is the culmination of the efforts of the committee, who spent much time preparing for the two conferences. Both were held at the Double Tree Downtown Hotel with the 2008 Forum from June 9th–12th, 2008 and the 2009 Forum from May 4th–7th, 2009. As you receive these proceedings, work is already underway on the 72nd Annual PEI Technical Forum, to be held in Nashville in May of 2010 and the upcoming International Enamellers Congress to be held in Chicago, tentatively in 2012.

First, I would like to thank everyone who attended the two conferences, and, in particular, those who participated by giving talks. I would also like to thank the members of the Technical Forum Planning Committee for planning and successfully executing these two conferences. We have all lived through stresses, strains, and sacrifices of the recent Great Recession. Our businesses are demanding more and more of less and less of us than ever. It is more difficult to get travel approved, hard to be away from the office and family, and it is more difficult to find the time outside normal business hours to do the work above and beyond required for an endeavor like the PEI. As such, everyone's extra effort is appreciated more than ever.

The PEI and the associated Back-to-Basics seminar have significantly educational and training value. Decades of experience and many viewpoints are brought together, and it is a unique opportunity to meet with multiple customers at once. The PEI Technical Forum is provides invaluable information and training to the industry for people of all knowledge levels. Marketers and design engineers worldwide learn of enamel and how to best put it to use as a low cost environmentally friendly finish capable for lasting for centuries. We need to continue fulfilling that purpose while training people new to the industry so it can be guaranteed a prosperous future.

This proceedings volume is dedicated to the memory of Dana Chase, Bob Long, Tony Mazuca, John Oliver, and Charlie Summers who all passed away during 2009.

Charles Baldwin
Ferro Corporation
2008–2009 Tech Forum Proceedings Chief Editor

2008 PEI Officers

Chairman of the Board
BOB HARRIS
Hanson Industries, Inc.

President
WILLIAM GANZER
Mapes and Sprowl Steel

Vice Presidents
CHUCK DOCKERY
Briggs Industries

MIKE HORTON
KMI Systems, Inc.

BILL MACDONALD
P G Bell, Inc.

NORM NAPIER
Shepherd Color Company

WARREN NORTON
Dofasco Steel

PETER VODAK
Engineered Storage Products

JEFF WRIGHT
Ferro Corporation

Executive Committee Members
JONATHAN DAVIES
Pemco Corporation

PETER DORITY
Coral Chemical Company

KEN KREEGER
Nordson Corporation

DON MCCORMICK
Electrolux Home Products

GLENN PFENDT
A. O. Smith Corporation

NICK SEDALIA
Whirlpool Corporation

Treasurer
KEVIN COURSIN
KMI Systems, Inc.

Executive Vice President
CULLEN HACKLER
PEI

2009 PEI Officers

Porcelain Enamel Institute
PO Box 920220, Norcross, GA 30010
Phone : 770-281-8980
E-mail : penamel@aol.com
www.buyporcelain.org / www.porcelainenamel.com

2008 Technical Forum Committee

2009 Technical Forum Committee

Chairman: Peter Vodak (Engineered Storage Products)
Vice Chairman: Mike Horton (KMI Systems, Inc.)

Jason Butz (Engineered Storage Products)
Jeff Dailidas (Nordson Corporation)
Peter Dority (Coral Chemical Company)
Holger Evele (Ferro Corporation)
Phil Flasher (ITW Gema)
Cullen Hackler (Porcelain Enamel Institute)
Mike Koenigstein (Roesch, Inc.)
Kara Kopplin (QTEC Consulting)
Ken Kreeger (Nordson Corporation)
David Latimer (Whirlpool Corporation)
Tony Mazzuca (Pemco Corporation)
Warren Norton (Arcelor Mittal Steel)
Liam O'Byrne (O'Byrne Consulting Services)
Tim Scott (Henkel Surface Technologies)
Lester Smith (Porcelain Consultants)
Larry Steele (Mapes & Sprowl Steel)
Jim Terwilliger (A O Smith Water Heaters)
Dave Thomas (American Trim)
Miles Votava (Ferro Corporation)
Jack Waggener (URS Corporation)
Mike Wilczynski (A O Smith Protective Coatings)
Ted Wolowicz (Electrolux Home Products)
Jeff Wright (Ferro Corporation)

Past Chairs of PEI Technical Forums

Holger Evele *Ferro Corporation*	2006–07
Steve Kilczewski *Pemco Corporation*	2004–05
Liam O'Byrne *AB&I Foundry*	2002–03
Jeff Sellins *Maytag Cooking Products*	2000–01
Robert Reese *Frigidaire Home Products*	1998–99
David Thomas *The Erie Ceramic Arts Company*	1996–97
Rusty Rarey *LTV Steel Company*	1994–95
Douglas Giese *GE Appliances*	1992–93
Anthony Mazzuca *Mobay Corporation*	1990–91
William McClure *Magic Chef*	1988–89
Larry Steele *Armco Steel*	1986–87
Donald Sauder *WCI–Range Division*	1984–85
James Quigley *Ferro Corporation*	1982–83
George Hughes *Vitreous Steel Products Company*	1980–81
Lester Smith *Porcelain Metals Corporation*	1978–79

Evan Oliver *Bethlehem Steel Corporation*	1977
Wayne Gasper *The Maytag Company*	1975–76
Donald Toland *U.S. Steel Corporation*	1973–74
Archie Farr *O. Hommel Company*	1971–72
Harold Wilson *Vitreous Steel Products Company*	1969–70
Forrest Nelson *A.O. Smith Corporation*	1967–68
Grant Miller *Ferro Corporation*	1965–66
Mel Gibbs *Inland Steel Company*	1963–64
Charles Kleinhans *Porcelain Metals Corporation*	1961–62
James Willis *Pemco Corporation*	1959–60
Lewis Farrow *Whirlpool Corporation*	1957–58
Gene Howe *Chicago Vitreous Corporation*	1955–56
W.H. "Red" Pfieffer *Frigidaire Division, G.M.C.*	1953–54
Roger Fellows *Chicago Vitreous Corporation*	1951–52
Glenn McIntyre *Ferro Corporation*	1948–50
Frank Hodek *General Porcelain Enameling and Mfg. Company*	1936–47

2008 PEI Tech Forum

DENSE PHASE PUMPING AND SPRAYING OF P/E POWDER

Ken Kreeger
Nordson Corporation

The presentation will review the development, advantages and use of dense phase pumping and spraying of powder porcelain enamel.

A review of the development , advantages and use of dense phase pumping and spraying of porcelain enamel powder.

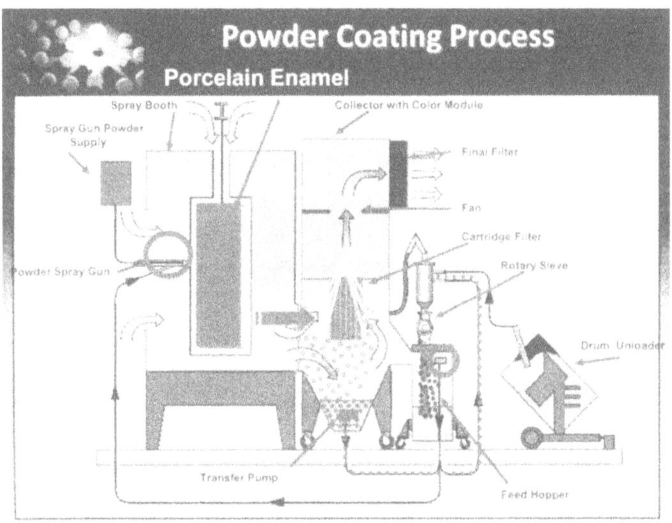

Traditional system with booth, recovery system, delivery system and guns.

How much lost production (profit on goods) is lost during downtime?

With faster color change and greater transfer efficiency, new “lean” non-reclaim systems are becoming popular.

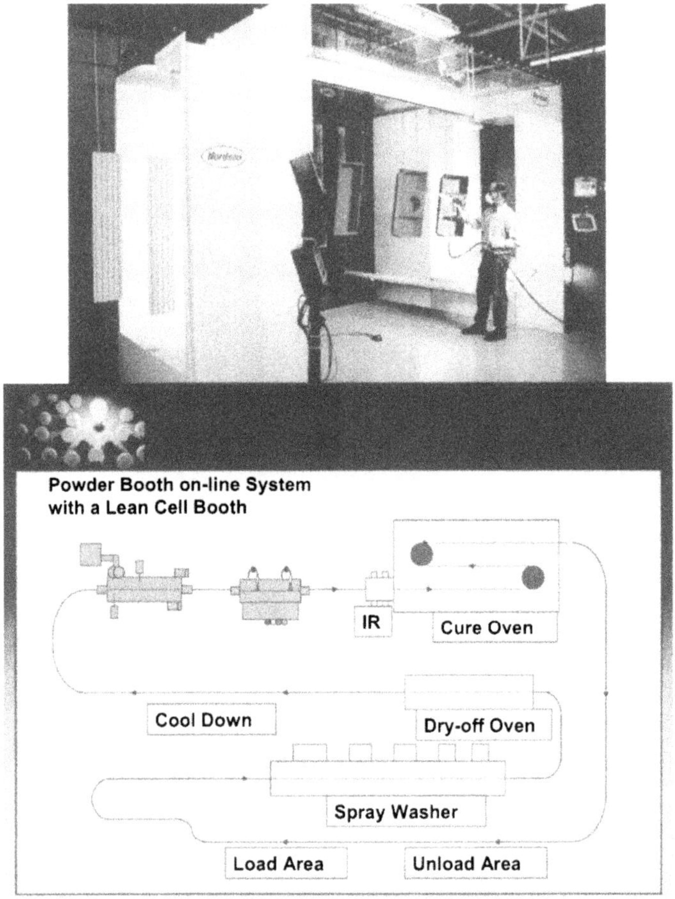

<u>Powder Coating Process</u>

- Powder coatings are fluidized in the feed hopper
- Fluidized powder is pumped to the spray gun
- An electrostatic charge is applied to the powder at the tip of the spray gun

- High voltage power supplies are used in corona charging systems to generate an electrostatic field.
- High voltage generators are integrated directly into the applicator.

Manual & Automatic Integrated Cascade producing 100,000 volts provides maximum charge at the gun tip.

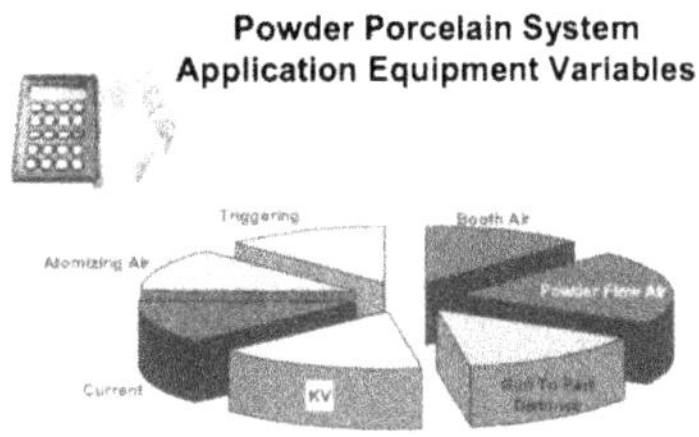

Historically most of the focus has been on electrostatics and charging system controls.
Put the pump can be a major contributor.

The powder pump is designed to deliver powder to the applicator in a controlled, consistent, and uniform flow.
This is typically accomplished through the use of Venturi:

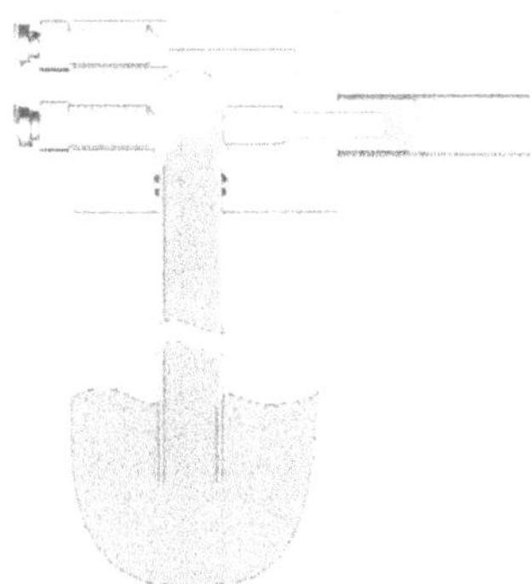

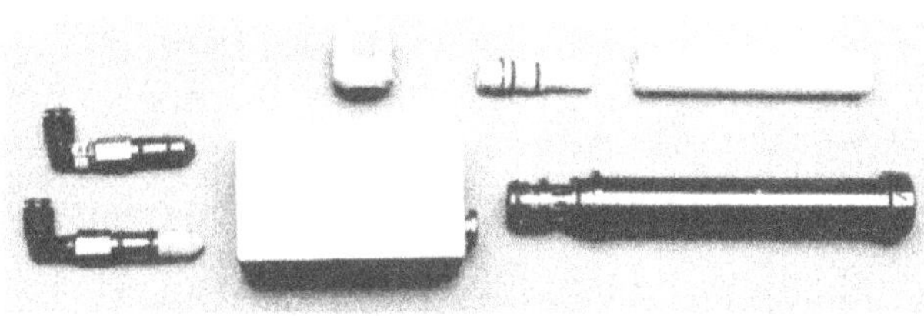

All powder-contact surfaces are made of wear-resistant ceramic materials

Aerodynamic play important role in powder application efficiency and coverage of recessed areas.

<u>We must address several issues.</u>

- spray pattern dynamics of an applicator.

- working characteristics of a powder pump.
- powder-to-air ratio.
- aerodynamic turbulence in the powder deposition area.

Prodigy Technology with HDLV
HDLV = High Density (Powder), Low Velocity (Air)

- Powder Hose has high concentration of powder, very little air
- Venturi pumps uses hi-velocity air – HDLV pumps meter powder using less air to propel

Dense Phase Pump Without Moving Parts

Designed to purge and color change quickly

See through design for easy diagnostics

Up to 27kg per hour output

Delivery hose has 6mm ID

At the centre of Dense Phase technology is the HDLV (High Density Low Velocity) pump.

It resembles a human heart in that it consists of two side by side pumps each with an upper and lower chamber working together to provide precise, carefully synchronized powder transport.

Spray pattern is formed independently of powder transport
Absolutely minimal air volume for powder delivery

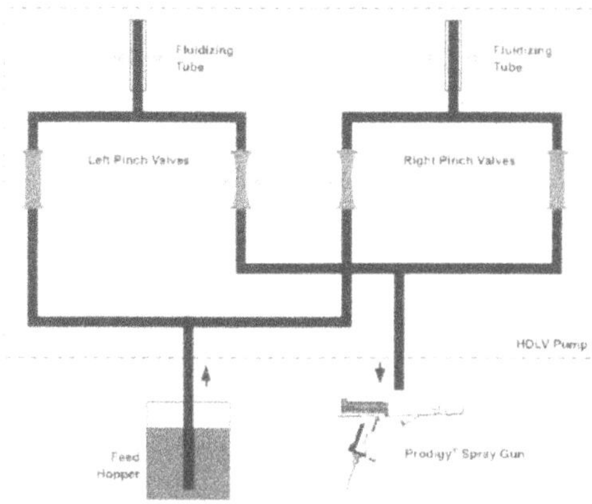

Conventional Powder Delivery (Low Density – High Velocity)

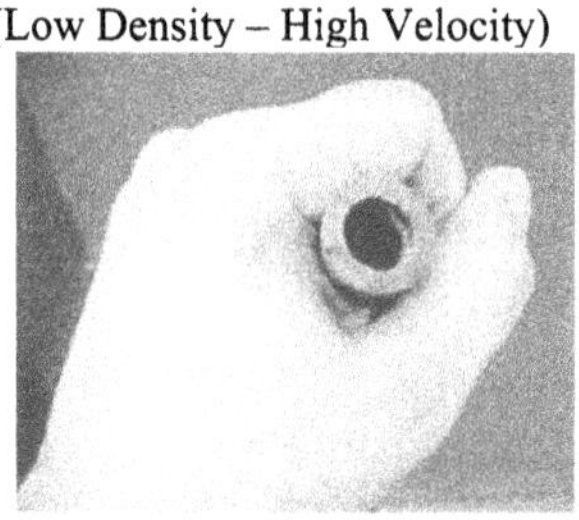

Powder Flow Rate 40 lb./hr

HDLV Powder Delivery (High Density – Low Velocity)

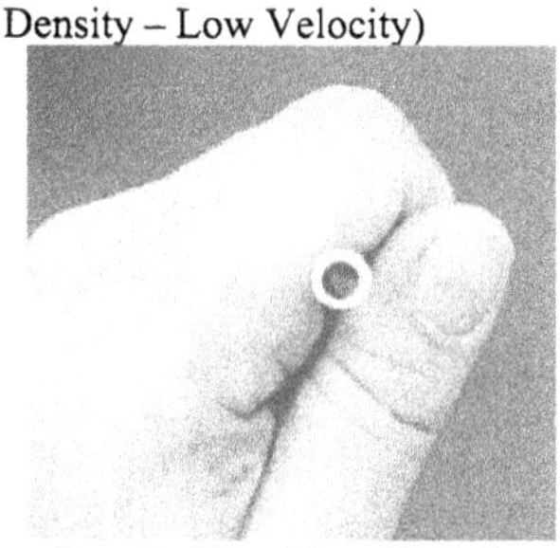

Powder Flow Rate 70 lb./hr

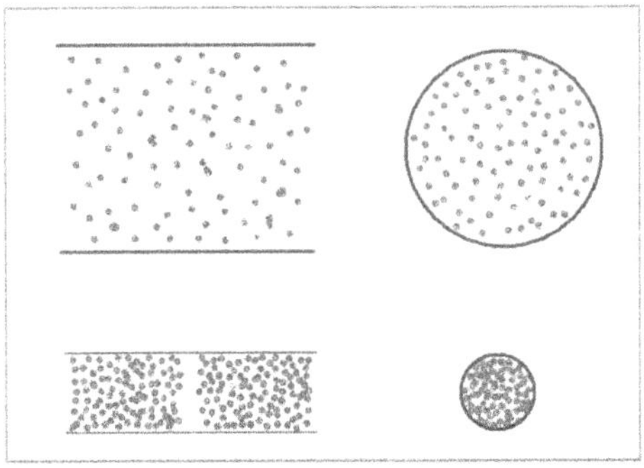

Stylish operator interface
Easy to use graphical interface with selectable language
AFC and Select Charge Functions
Colour change button purges system
Easy to use LCD Graphical Digital Display
System designed to Purge and Color Change Quickly
- One button color change
Before
Sure Coat guns
Conventional booths
Line speed 12 fpm
Low flow application – Low TE

After
Prodigy guns
New Lean Cell Systems
Line Speed 20 fpm
Low flow application – High TE

Drum unloader – virgin powder delivery

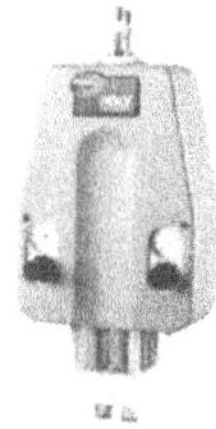

Simple pneumatic control for ease of operation
Less compressed air used to propel the powder results in more efficient recycling
Transports large amounts of powder in short amount of time
Automatically purge-able with either continuous or pulse purge

DEVELOPMENT OF THE COLD ROLL RP45 GRADE FOR WATER HEATER TANK BODIES

Stan Lipkowski
ArcelorMittal Steel

The paper will provide an overview of the ArcelorMittal development of a cold roll steel grade for water heater tanks that has good strength retention after enamel firing.

An overview of the development of a cold rolled steel grade for water heater tanmks that has good strength retention after enamel firing.

Outline

- ArcelorMittal – Who We Are
  - Formation of ArcelorMittal
  - Company Information

- Global R&D
  - World Wide Research Centers
  - Ghent Enameling Laboratories
  - Recent R&D Development: Warping Simulation

- RP45 Grade Development
  - New Product Development Process
  - Tank Design Aspects
  - Grade Strengthening Mechanisms
  - Typical Weight Savings

ArcelorMittal
The World's Leading Steel Company

- ArcelorMittal is the number one steel company in the world with revenues of US$ 105.2 billion in 2007 with shipments of 109.7 MT, representing around 10% of world steel output.
- The company employs ~310,000 employees in more than 60 countries.
- The company has an industrial presence in 27 countries across Europe, the Americas, Asia and Africa.
- The company holds large captive supplies of raw materials and leadership positions in all major global customer segments: automotive, construction, household appliances, and packaging.
- Our Vision: "To be the World's Most Admired Steel Company"

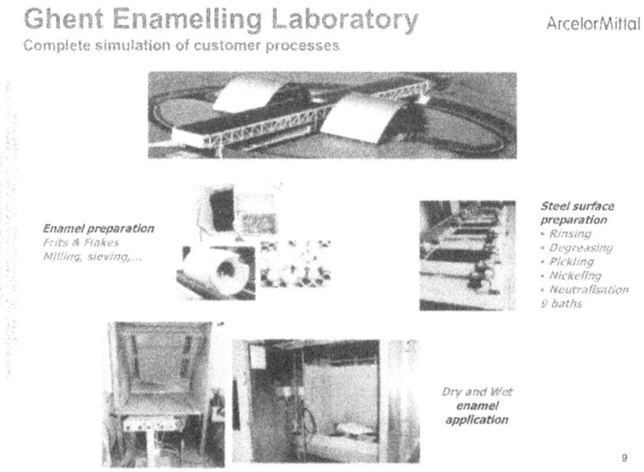

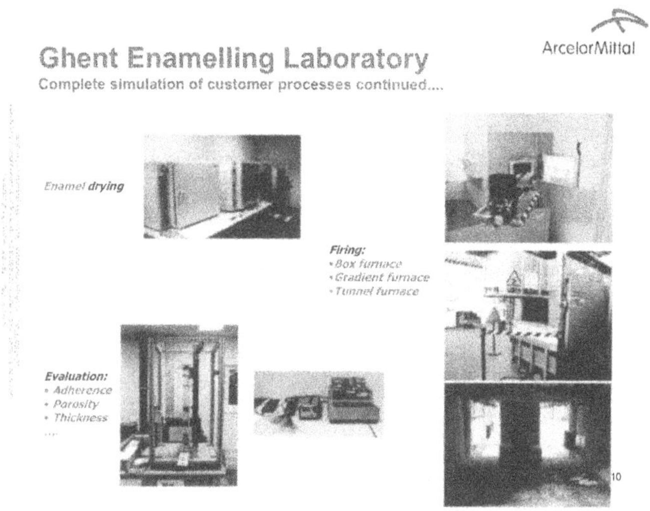

Recent R&D Development:
FEA Warping Simulation

- Warping of (enamelled) panels/assemblies
    - After enamelling cycle , accounting for
        - Enamel type

RP45 Grade Development

- Product developed for the Water Heater Industry
  - Replaces hot roll steel in water heater tank bodies
  - Higher strength 45ksi min. product
  - Allows customers to become more cost competitive
  - Commercialized in 2006

New Product Development Process

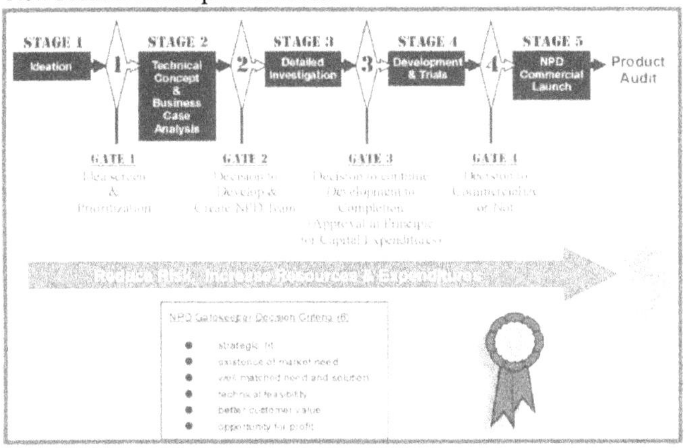

Tank Design Aspect: Buckling Analysis
Design equation for max. internal pressure:

$$p_0/\gamma F = \frac{80(r/D)^{0.825}}{(D/t)^{1.5}(R_s/D)^{1.15}}$$

» p_0: allowable internal design pressure (MPa)
» γ: factor for strain hardening
» F: 0.2 percent proof stress of the as-received material (MPa)
» t: *minimum* thickness in knuckle region (mm)
» r, R_s and D as shown in figure (mm)

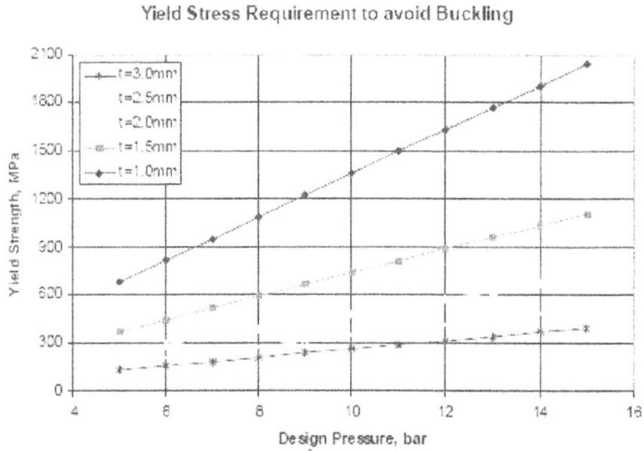

- To avoid buckling due to internal pressure, increasing yield strength is necessary for lower thicknesses
- More accurate calculations are needed once a specific geometry is known

Tank Design Aspect: Von Mises Stress

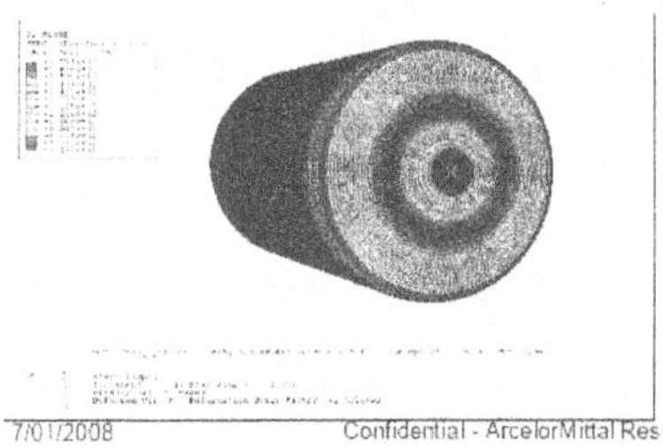

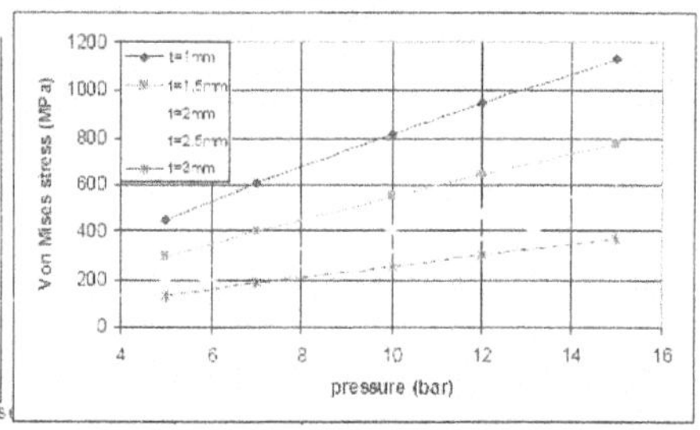

RP45 Grade Strengthening Mechanisms

1. Grain Size Control (Fine grain microstructure): Hot Mill Coiling Temperature and Batch Annealing Cycle
2. Chemistry: Higher Phosphorus, Manganese and Silicon levels

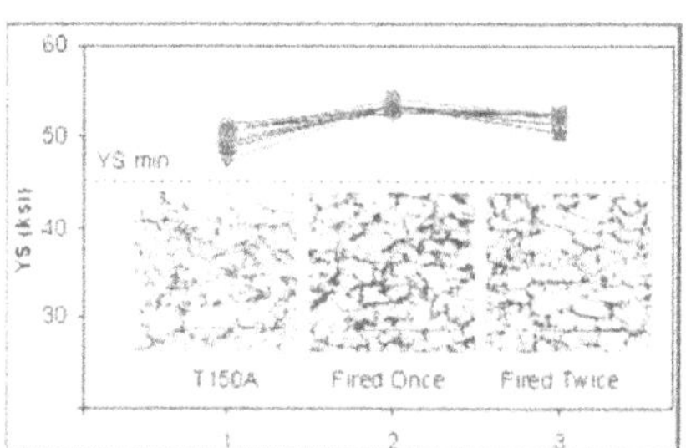

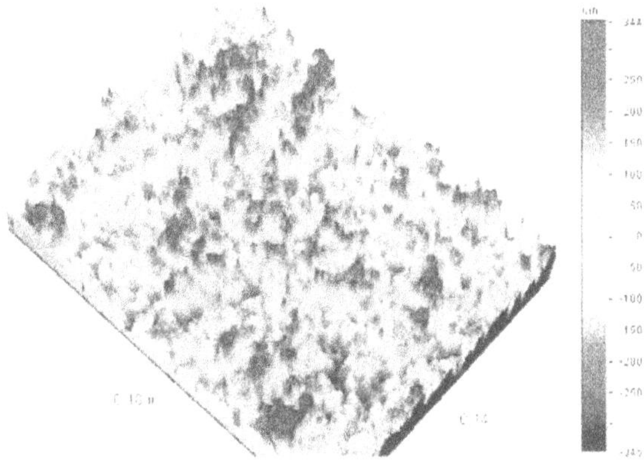

- Obtained by rolling with specially roughened rolls at the cold mill and temper mill
- Surface does not require shot blasting for enamel adhesion

ArcelorMittal

Typical Weight Savings

- Typical water heater tank body weight savings achieved by converting from hot roll to cold roll RP45

Tank Diameter	Weight Savings
14"	33%
16"	22%
18"	21%
20"	17%

19

Conclusions

- To reduce the thickness of water heater tank bodies, higher yield strength material is required
- By converting water heater tank bodies from hot roll steel to the higher strength cold roll RP45 product, customers can achieve:
    - 20% weight savings on average
    - Elimination of the shot blasting operation
    - Significant cost savings

ENERGY COST IN WASHER SYSTEMS

Brian Rozdilsky
KMI Systems, Inc.

This is a brief overview of where the costs are in a washer system and an exploration of ways to mange/control/reduce those costs is discussed.

Topics:
- Areas for Energy Improvement
- Energy Savings
- Cost Savings
- Closer Look

Areas of potential improvement:
- Areas for Energy Improvement
- Energy Savings
- Cost Savings
- Closer Look

Where does the energy go?

REQUIRED PROCESS TEMP
- Drop 10 °F from 160F to 150F
  - Saves 550,000 BTU/HR
- Drop 10 °F from 150F to 140F
  - Saves 400,000 BTU/HR
- Drop 10 °F from 140F to 130F
  - Saves 290,000 BTU/HR

Insulate Tanks
- Insulating tanks can save 26,000 BTU/HR

Upgrade to High Efficiency Motors

- Saves 140 KW/Year

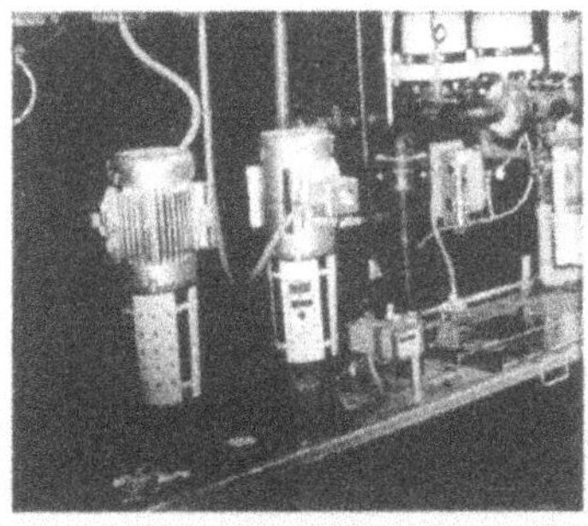

Adding VFD will reduce energy usage by 8,614 KW per motor.

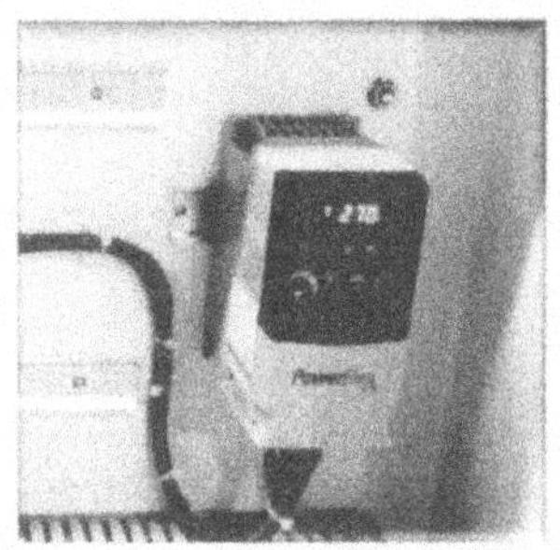

Is your Wash System Over Sized

- Reducing Mask Opening Can Save 100,000 BTU/Hr
- Sized for Speed
- Sized for Part Length

Conclusions

- Lower Process Temp 10°F
    - $8,000 to $14,000 per year per stage
- Insulate Tanks
    - $ 561 per year per stage
- High Efficiency Motors
    - Saves $125 per year per motor
- Install VFD's on Pumps
    - $ 603 per year per pump
- Stream Line Washer Design
    - $2,000 to $ 13,000 per year

COMPARISON OF COBALT AND NICKEL IN FRIT FORMULATIONS

Mike Wilczynski, Roger Wallace
A.O. Smith Corporation
Protective Coatings Division

The main adherence agents in porcelain enamel are oxides of nickel and cobalt. This laboratory study presents a comparison of these two oxides and their relative effectiveness.

The purpose and scope of this paper is to determine the relative effectiveness of Cobalt Oxide versus Nickel Oxide as bonding agents in frit formulations. There are two main questions that need to be answered. First, is there a ratio of how much Cobalt to add versus how much Nickel to add to a frit to get the same bond results? And second, given the current price levels of these two materials, which is more cost effective?

We chose two types of frit formulations for our experiment. A water heater (wet spray applied) groundcoat frit, and an electrostatic powder groundcoat frit. The level of cobalt oxide and Nickel oxide was varied in each frit. The bond levels were then compared.

The water heater groundcoat frit was milled in a typical wet mill addition. Firing temperature was 1580 degrees F. Steel surface was a light coil blast.

Cobalt and Nickel levels were varied as in the chart below:

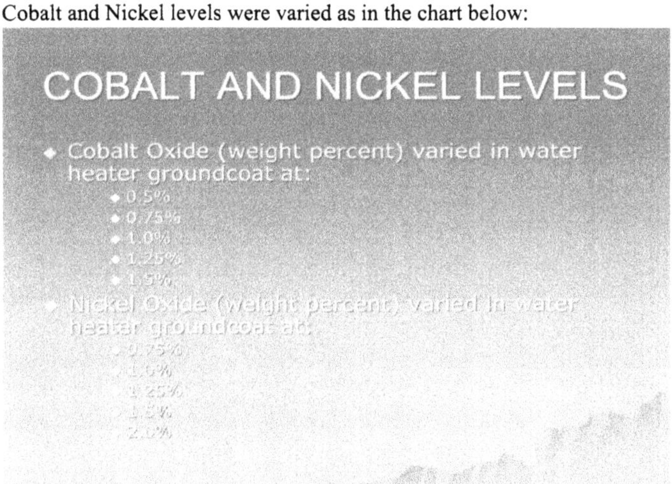

Subsequent bond ratings were checked, and charted in graph form:

A smaller amount of Cobalt Oxide was needed to generate the same bond level than a larger amount of Nickel Oxide. About twice as much Nickel Oxide in the frit generated the same bond as Cobalt Oxide.

The second experimental frit was an electrostatic powder groundcoat, dry milled, and fired at 1500 degrees F on clean-only cold rolled steel.

Cobalt and Nickel levels were varied as in the chart below:

Subsequent bond ratings were checked, and charted in graph form:

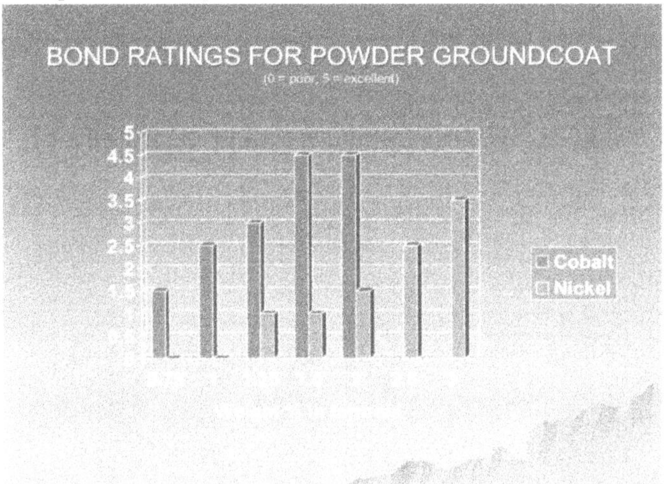

Again, a smaller amount of Cobalt Oxide was needed to generate the same bond level than a larger amount of Nickel Oxide. About two to three times as much Nickel Oxide in the frit generated the same bond as Cobalt Oxide.

Estimated Nickel to Cobalt bond ratios in the frits were as follows:

A historical price chart (from 2006 to present) of Cobalt versus Nickel LME prices will show that, at the ratios described above, using Cobalt as a bonding agent was cost effective until the middle of 2007. After that, the price of Cobalt diverged sufficiently from the price of Nickel so that the use of Cobalt Oxide as a bonding agent became less coss effective than using Nickel Oxide, as the price ratio between Cobalt and Nickel is now on the order of four or five to one.

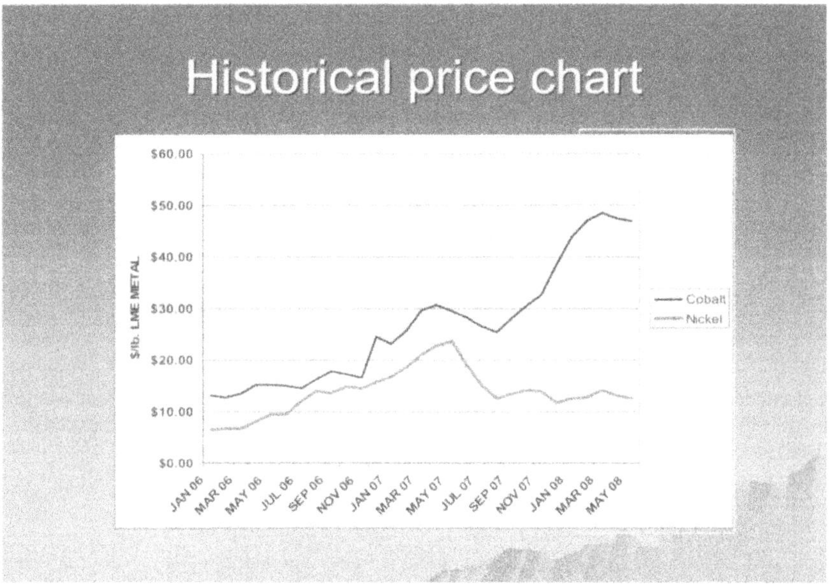

However, there are a few final observations and conclusions we would like to make. They are summarized as follows:

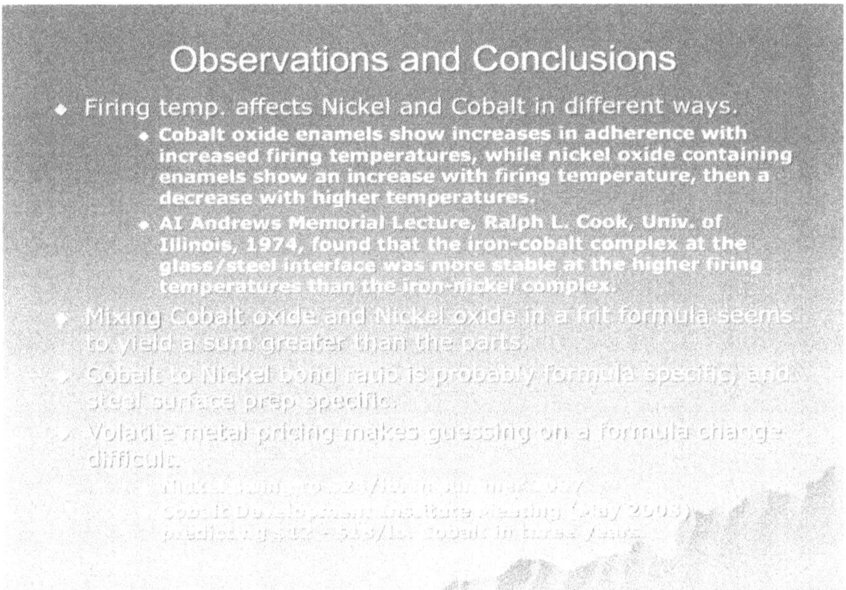

RAW MATERIALS AND ENERGY: INFLUENCE ON ENAMELS MARKET

Bruce Hanna
Pemco Corporation

In this paper the influence of raw materials and energy are considered in an examination of the porcelain enamel marketplace.

Summary:

- Introduction
- Market Pressures
- Raw Materials
  - Lithium
  - Nickel
  - Cobalt

- Possible developments on enamel

Developing Countries & New Markets
Consumption Increase

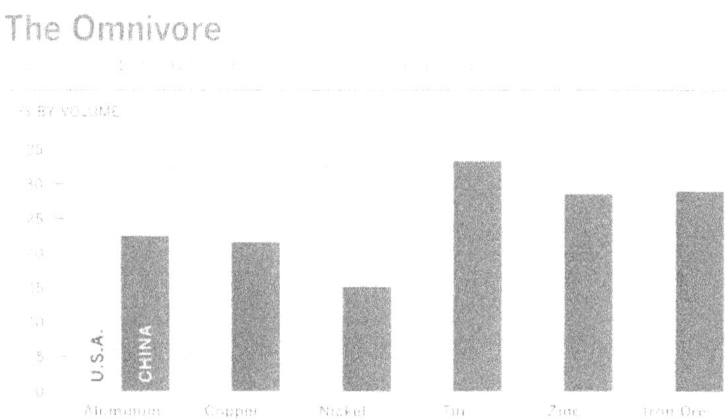

Consumption Increase New Markets

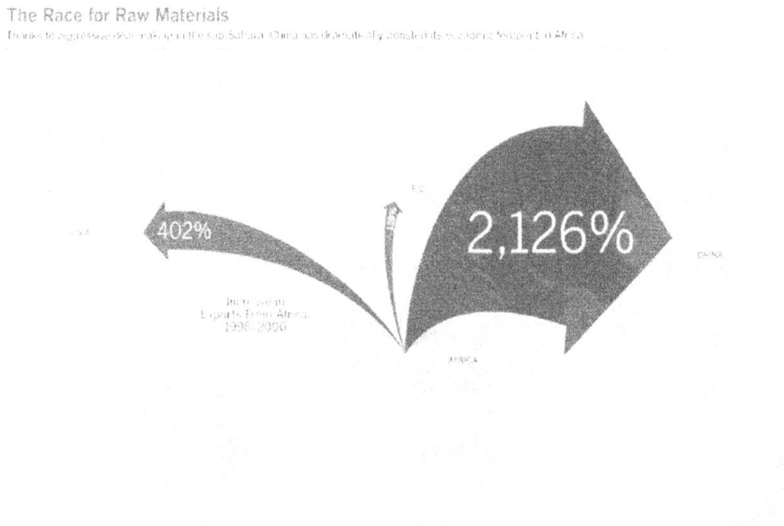
The Race for Raw Materials
402%
2,126%

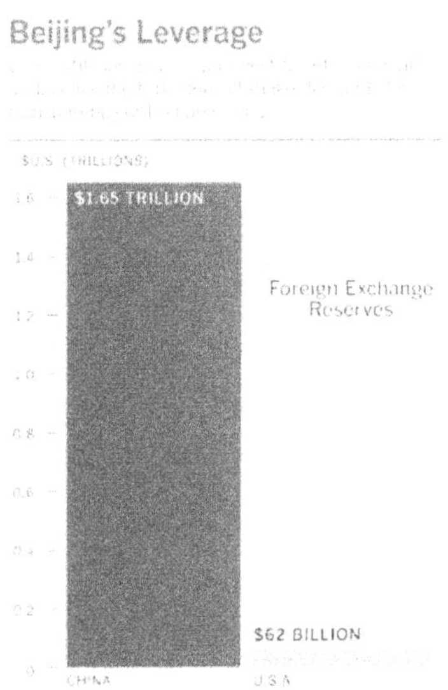
Beijing's Leverage
$1.65 TRILLION
Foreign Exchange
Reserves
$62 BILLION

The Lure of Copper

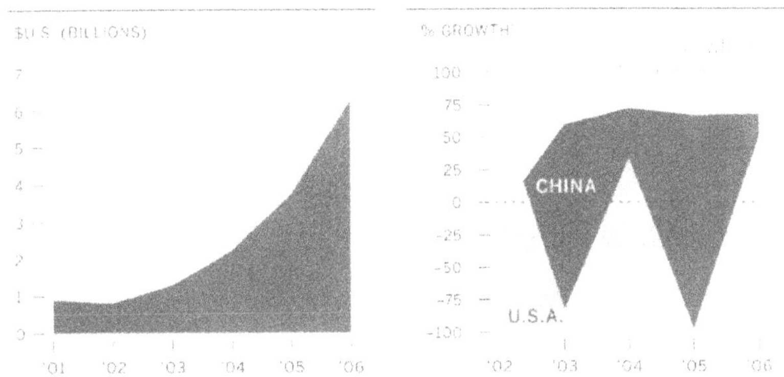

Increase Transport & Oil

A Black Hole for Black Gold

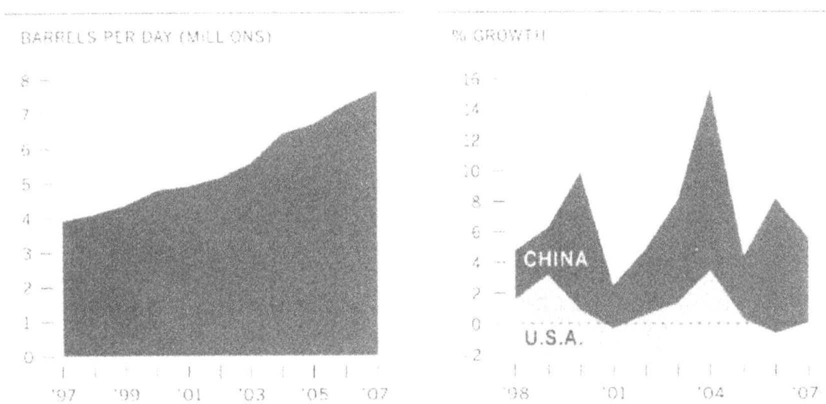

ELECTRICITY
Price linked to petroil
Energetic policy of the country
METHANE/NATURAL GAS
Relationship with suppliers countries
Consumption for enamel production and use

Lithium: alcaline element at low molecular weight
Production sources
BRINES FROM SALT LAKES

Production coutries
1) Chile (50% growing)
2) Argentina (growing)
3) US (stable)
Future developments
- China

HARD ROCK ORES MINES
Production countries
1) Australia
2) China
3) Canada
4) Zimbabwe
5) Democratic Republic of Congo (DRC)
Future developments
- Finland (new process)
- Canada

World consumption (as Li)
14.100 ton/year (official data 2003)
17.000 ton/year (estimation 2005)
Average growth 4 – 5% year

Price trend
1998: price slow down, chilean suppliers entrance
1998 - 2004: relatively stable prices
2004 - 2007: esponential growth

Nickel: "rare" metallic element
Productive sources (2004 data)
SULPHIDE ORES (58%)

LATERITIC SOILS
Limonite (28%)
Saprolite (14%)
Actual estimeted world reserves
62 mio ton
Mines sited in:
1) Australia
2) Russia
3) Cuba
4) Canada
5) Brasil
6) others
World consumption (2005)
Metallic alloys (79%) about which
Austenitic stainless steel (60%)

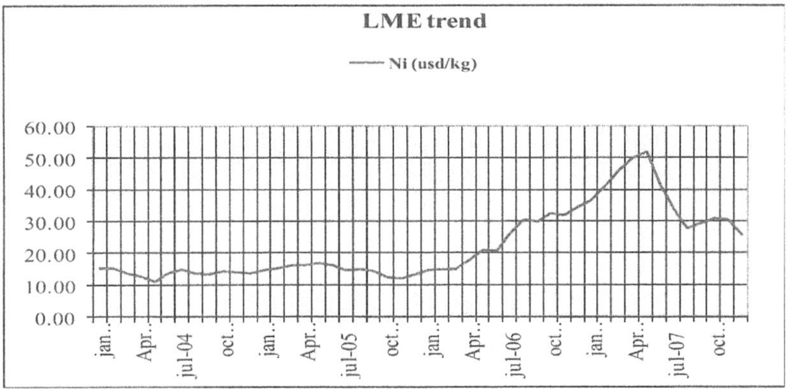

Cobalt: "rare" metallic element
Sub-product of extraction process of:
Nickel
Copper
Other metals
Actual world reserves
15 mio ton

LUSTREFLEX COATINGS FOR PORCELAIN ENAMEL

E.A. Axtel III and Klaus-Deiter Fritsche
Ferro Corporation

For many years glass coatings have served to brighten and beautify the world. Myriad soft drink and beer bottles have been coated, allowing customers to recognize their favorite brands immediately. Recent building designs are incorporating more coated glass with both color and functional properties. The design of appliances should be no different. New designs and decorating techniques will imbue appliances with novel appearances allowing consumers new choices when contemplating appliance purchases.

Outline
- Introduction to Functional Coatings
- LustReflex Coatings on Porcelain Enamel
  - Decoration
  - Stainless-Steel-Effect
- Conclusions

Introduction
- A new area has emerged in Glass Research at Ferro: Functional Coatings
- Research began in this area began with our work on the super-hydrophobic Lotus Effect coating for glass

Lotus "Spin-offs
- Success with the Lotus Effect coatings encouraged attempts to expand the chemistry of the coating in attempts to increase durability
- Other surface properties were observed with continuing work

Characteristics of Functional Coatings
- Pattern-able and spray-able coatings composed of dissolved metal ions
- Application of these coatings to glass and other vitreous substrates (like PE) provides new surface properties after firing

- Coatings are are theoretically capable of providing a wide array of functionalities to vitreous substrates
  - Color
  - Reflectivity
  - Conductivity
  - Scratch Resistance

Color and Glass Performance Materials

Shade Band Coatings

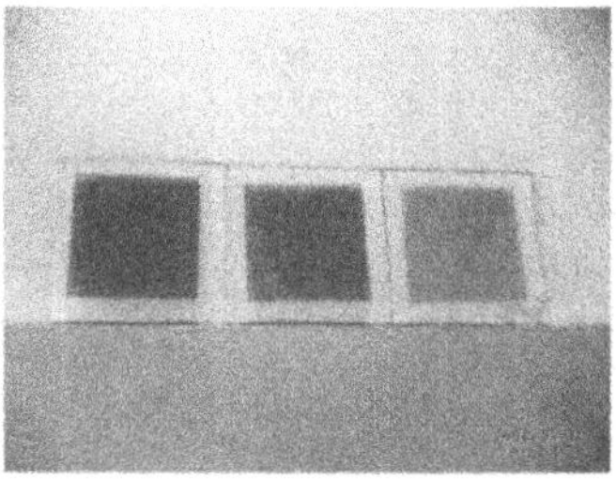

Color and Glass Performance Materials

LustReflex on a Building Façade

LustReflex Coatings

- Can be applied on a vitreous surface to provide a semi-mirror appearance
- Easily patterned by design of the screen
- Can be formulated as a spray for complex parts
- The basic reflective color is silver
- Interference colors are possible

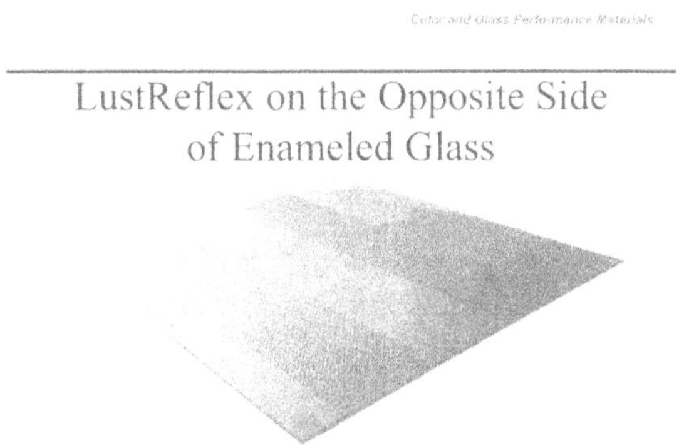

Imitation Brushed Stainless Steel Effect on Glass

- The coating is applied by a multi-step screen printing and firing method
- The coatings are fired under normal glass processing conditions

Imitation Stainless Steel Effect on PE
Step 1 - Start with a fired gray enamel on a metal substrate
Step 2 - Use a tungsten carbide file to lightly "brush" or scratch the surface of the PE
Step 3

- Spray LustReflex Primer and fire
- Spray LustReflex Coating
- Fire the coating in a normal processing fire

Conclusions

- The LustReflex coating belongs to the new family of products called Functional Coatings for Glass
- The product is a semi-mirror coating that can be applied to vitreous surfaces
- The LustReflex coating has been used to create an Imitation Stainless Steel Coating for both Glass and PE

BULK ENAMEL DISTRIBUTION SYSTEM

Jeff Walker and Steve Sloan
A.O. Smith Water Products – Johnson City Plant

A discussion on the benefits of an enamel distribution system that uses bulk storage tanks and a piping system to deliver enamel slip to each spray booth instead of individual pressure pots.

60 Gallon Pressure Pots

Numerous pots requiring numerous repairs and often working in an inconsistent manner.

Centralized bulk handling system on one line.

Company Profile

- Manufacture Residential & light Commercial, Gas & Electric Water Heaters
- Sizes from 20 to 119 Gallon Capacity
- Located Northeast Tennessee, Foothills of the Appalachian Mountains
- Approaching 6 million man-hours without a lost-time injury

- Tank Plant
  - Fabricates internal welded, glass-lined tank
  - 2 Enamel Depts. supporting 3 Tank Fabrication Lines
  - (7) Enamel Spray Booths
  - (2) KMI Furnaces
  - Process from 1600 to 2000 gallons slip per day
  - (2) Bulk Distribution Systems
    - 1st Bulk Distribution System(s) >25 years old

- **Shell Spray Booth**

- **Flue Spray Booth**

- **Bulk Enamel Distribution System**
- System Description
    - Piping
        - Circulating Closed Loop from Bulk Tank to Spray Booth(s) & Return Line
        - 2" Galvanized Pipe
        - Minimize Distance from Tank to Spray Booth(s)
        - Minimize Fittings, Directional Changes, i.e. 90° elbows, unions, reducers, etc.
        - Long Sweeping Elbows
        - Gasketed Flange Connections, No Unions
        - Full Flow Ball Valves
        - Pee Valves
        - Water Supply Plumbed at each Circulating Pump

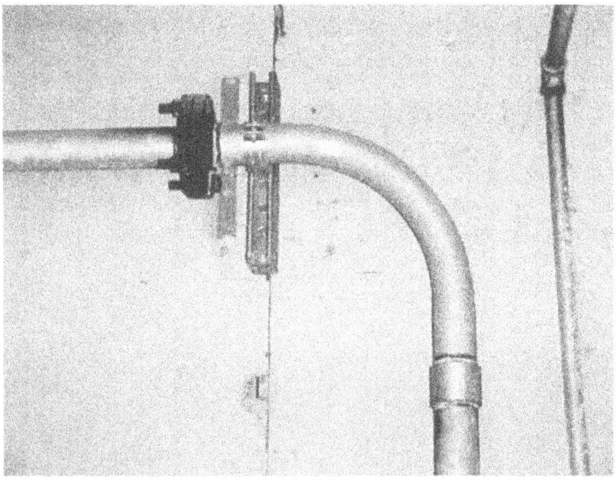

- **Sweeping Elbow**

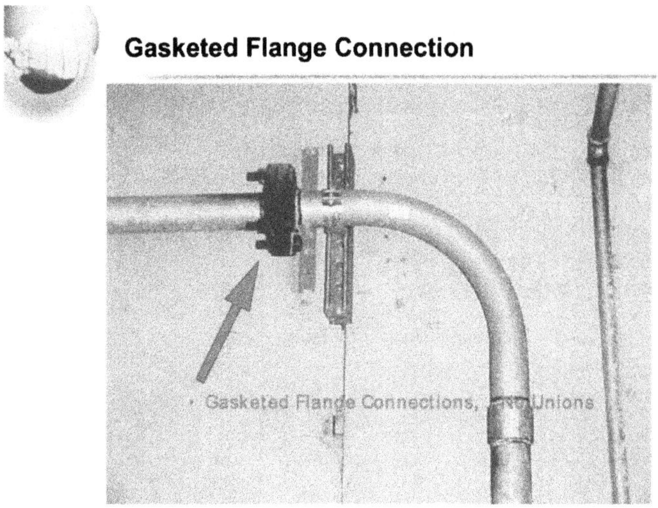

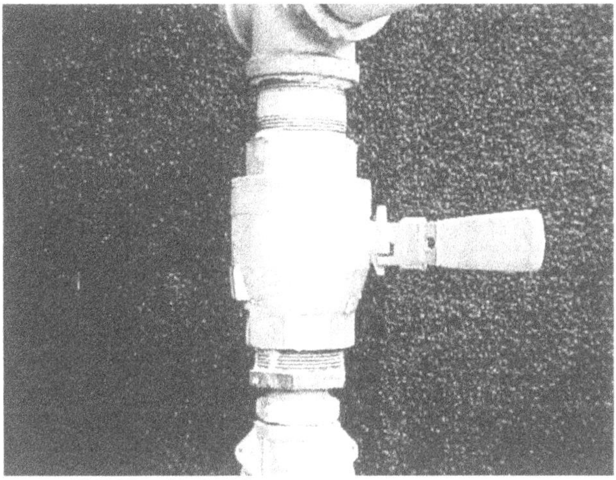

- **Full Flow Ball Valve**

- **"Pee" Valve**

- **Circulating Diaphragm Pumps**

Tranquilizers

- Enamel Characteristics / Application
  - Milled Groundcoat
  - Fired Thickness = 9 mils, ±2
  - Specific Gravity = 1.81-1.83, (No Change)
  - Fineness = 10-13, (No Change)
  - Slump = 5-1/2 to 5-3/4
    (Changed, was 5-3/8 to 5-1/2 w/ pressure pots)
- Initial Problems / Concerns
  - Enamel Buildup In Lines - (Lines Flushed Nightly)
  - Risk of Contamination – (Happens but rarely)
  - Squeeze Valve Life – (Replaced w/ "pee" valve)
  - Pump Failure – (Rebuilt at 8-12 months)
  - Tranquilizer Failure – (1/5 replaced in 5 years)
  - Ball Valve & Pipe Wear – (Ball valves replaced often)

Conclusions:

- Pressure Pots
    - No Longer Fill Individual Pressure Pots
    - No Longer Setup Individual Pressure Pots
    - No More Maintenance on Pressure Pots
    - Reduced Material Handling
    - Reduced Overtime
    - Better Material Utilization
- Benefits
    - Improved Quality
        - Operators No Longer Make Constant Gun Adjustments Throughout the Day
        - Process Control Much Better
        - Smoother, More Consistent Spray
        - More Consistent Coating Thickness
    - Reduced Operating Costs
    - Increased Productivity
        - Eliminated Downtime Changing Over Pressure Pots
        - Reduce number of adjustments required through the day

PORCELAIN ENAMEL RECLAIM TECHNOLOGY REVIEW

Mike Horton
KMI Systems

There are many types of enamel reclaim systems in use today. This paper will review the various technologies and discuss their advantages and disadvantages.

What is porcelain enamel reclaim?
The process of recycling enamel slip or powder from coating operations to beneficially apply to surfaces.

What are the benefits?

- Reduce Coating Material Cost
- Less Solid Waste Disposal Cost
- Reduced Water Treatment Cost
- Less Energy & Labor to Prepare the Coating

What are the disadvantages?

- Requires More Labor and Energy
- Additional Equipment is Needed
- The Plant Environment Must Be Controlled
- An Additional Coating Process is Needed
- Rejects May Increase, Resulting in Increased Cost, Scrap and Reduced Production

Reclaim Process

Processes	**Reclaim**
Mill Slip:	Pit Wash Down
Pre Mill Slip:	Less Wash Down
Screen + Magnetic Separate Pump to Storage Wash Down Tanks, Mills Floor, Screens and Lab:	Pit
Spray 30-60% Transfer Efficiency:	Reclaim 85%
Flow Coat Wet or Dip 90-100% Transfer Efficiency:	Reclaim 95%
Powder Frit Coat 40% Transfer Efficiency	Reclaim 95%
Brush, Bead and Drain, Blow-Off:	Reclaim

Tool Washer:	Reclaim Wet
Powder Hanger Cleaner:	Reclaim Dry or Wet
Dust Collector:	Reclaim Wet
Spray Washer:	Pit/Water Treatment
Wash Down:	Pit Collection

Baffles Drop 85-90% of Overspray

Overspray Reclaim Wet Process Collection:

Dust Collector:	10-15% to Collector
Tool Washer:	Centrifugal Separate and Settle
Powder Hanger Cleaner:	Pump to Booth Screen or Collect and Mix with Wet Enamel

DUST COLLECTOR RECLAIM

Hanger Cleaner

Reclaim Process

Pit Collection	Frit Settles
Mill Room or	Mechanical Agitate
Process Area:	Air Agitate with Compressed Air or Blower

Reclaim Applications

Booth Over Sprays:	Reuse on High Appearance Surfaces
Dust Collection Materials:	Mix with Wet Ground Coat Backside of Parts
Tool Cleaner Materials:	Wet Ground Coat on Small Parts
Pit Materials:	Backside Tubs Water Tank Small Low Appearance Parts

Conclusions

- With raw materials and energy cost, a successful reclaim process must be established and maintained.
- Wet enamel oversprays can be recycled when sprayed in controlled booth environment.
- Frit Materials can be collected from tool cleaners and pits and reused on low appearance surfaces.
- Each enamel process will require a custom design reclaim process and control procedure.

PROGRESS IN THE USE OF SPECIALIZED ENAMEL COATINGS TO BOND CONCRETE TO STEEL

Philip Malone, Sean W. Morefield, Charles A. Weiss and Earl Baugher
U.S. Army Corps of Engineers
Engineer Research and Development Center
Vicksburg, MS and Champaign, IL

Cullen L. Hackler
Porcelain Enamel Institute (PEI)

An update on development efforts in joint work by the U.S. Army Corps of Engineers and the Porcelain Enamel Institute

Outline

- Review of technology
- Benefits we can expect
- Status of Implementation of Technology
  - Research/Demonstration Activity
    - Validation of Bonding Approach
    - Hardened Structures
    - Marine Construction
    - Investigation of Self-Healing of Bonding Enamel
- Applications
- Technology Transfer
- Summary

Problems with Using Steel Reinforcement in Concrete

- The concrete-to-steel bond is the weak part of the composite
- Corroding steel is increasing in volume and cracks the surrounding concrete

I-95 Bridge

Porcelain enameled reinforcing steel (PERS) can solve both problems

Basic Technology

- Tricalcium silicate (C3S) can be mixed with an alkali-resistant enameling frit
- The frit can be fired onto reinforcing steel without changing the hydrating properties of the C3S
- When the coated steel reinforcement is embedded in concrete, the enamel lets the steel bond to the surrounding concrete

Increase in Bond Strength

Comparison Of Average Bond Strengths

Treatment	Average Peak Force lbf (N)	Std. Deviation lbf (N)	Average Bond Strength psi (MPa)
Steel fiber embedded in mortar7	---	---	295.4 – 394.5 (2.04 - 2.72)
Steel rods, uncoated embedded in mortar	588.7 (2,618.2)	104.8 (466.2)	298.8 (2.06)
Enameled rods without portland cement embedded in mortar	786.4 (3,497.9)	121.6 (540.8)	391.6 (2.70)
Rods with enamel containing portland cement embedded in mortar	2500.9 (11,124.6)	52.9 (235.3)	1274.9 (8.79)

Protection from Corrosion

- On exposure of 3.5% NaCl solution corrosion does not occur unless the enamel coating is removed
- Vitreous enamel does not delaminate like organic coatings do
- Vitreous enamel is the most durable coating that can be put on steel

- Post-fabrication with good quality control could decrease failure from reinforcement corrosion

Current R&D --Validation

- MS&T duplication of enamel rod tests embedded in mortar
- Result were in agreement with tests that were independently conducted at ERDC
- Enameling does not appear to change the properties of the steel

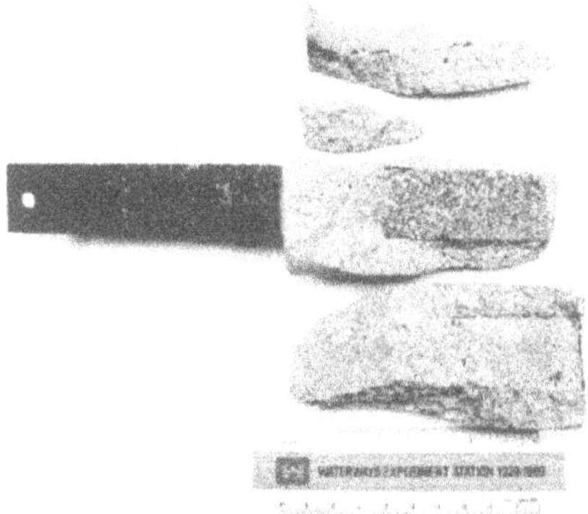

Tests with composite enamel-coated straps demonstrated strong

Current R&D -- Structural Hardening

- Earthquake and blast protection are serious concerns
- Research projects underway at MS&T relate to blast resistant barriers
- Proposals on projectile penetration have been submitted

Failure of reinforced concrete column in compression

Current R&D -- Marine Construction

- USAE Construction Engineering Research Laboratory has DoD funding for marine construction demonstration
- Project will replace failing reinforced concrete platforms in use at U.S. Army Depot Corpus Christi

Chloride invasion causes rebar corrosion, concrete spalls and structure fails

Current R&D – Self-healing Enamel (1)

- Portland cement grains hydrate in the enamel as soon as they are in contact with moisture
- Hydration proceeds with formation of gel and filling of space between grains
- Cement covered in glass does not hydrate until exposed to moisture by cracking

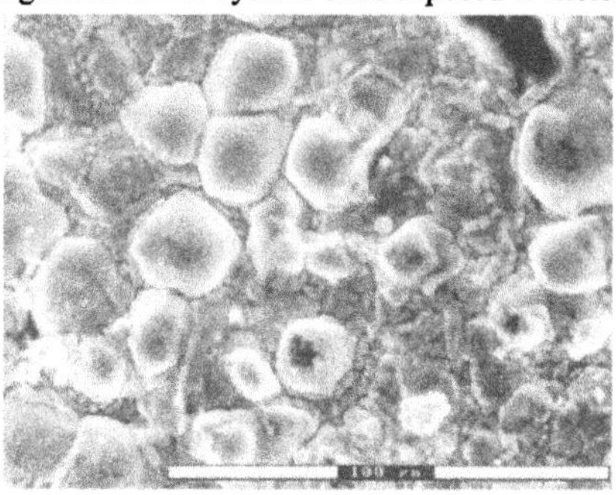

Hydrating cement grains on surface of enamel

- Scanning electron microscopy is being used to investigate self-healing effects in the C3S-vitreous enamel composites
- The increased local alkalinity can passivate exposed steel

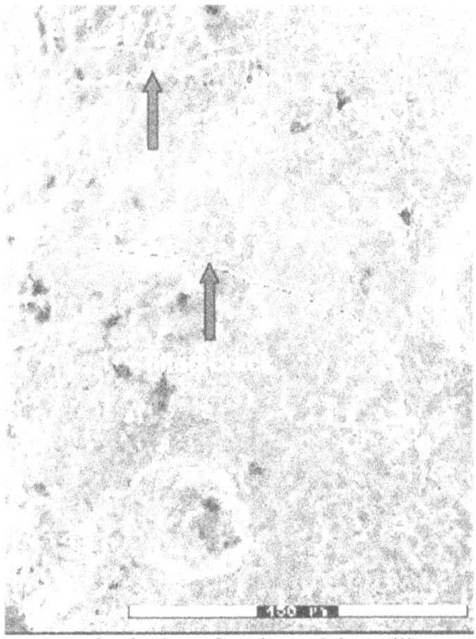

Embedded cement grains hydrate forming calcium silicate gel in fractures

Presentations and Publications on Enamel in Bonding Reinforcement

- 2nd Symposium On Connections between Steel and Concrete, Stuttgart, Sept. 4-7, 2007
- CORROSION 2008, National Association of Corrosion Engineers, New Orleans, March 16-20, 2008
- 2008 Concrete Bridge Conference, St. Louis, MO, May 4-7, 2008
- 21st International Enamellers Congress, Shanghai, May 18-22, 2008
- MEGARUST 2008, Marine Coatings and Corrosion Conference, Louisville, KY, August 18-21, 2008
- Excellence in Concrete Construction through Innovation, Kingston University London, September, 9-10, 2008
- FACT SHEET on Enamel for Protection and Bonding of Reinforcing Steel has been prepared
- MS&T has reproduced and validated the bonding data (consistently obtaining 3-4X increase in bond strength)
- Support has been obtained for development of new protective structure using bonding enamel technology
- DoD Corrosion Program has approved a demonstration project (Army Depot Corpus Christi, TX
- Technology transfer program is underway
- Additional applications for the technology are appearing

PERFORMANCE COATINGS FOR ENERGY EFFICIENCY

Holger Evele, Charles Baldwin, Terry Detrie and Dan Swiler
Ferro Corporation

Today's consumers want to reduce their energy usage. Coatings with novel release properties can simplify cooking and the eventual cleaning afterward, while infrared-reflective coatings offer better insulated buildings and more efficient cooking.

Introduction

Two options for improving energy efficiency using porcelain enamels were evaluated. First, the infrared reflectivity of the coating could be tuned to either keep areas cool to reduce climate control usage or increase cooking efficiency. Second, special coating formulations could reduce energy usage.

Part 1 – Infrared Reflectivity

The ability to engineer the infrared reflectivity of an article creates significant opportunities to improve energy efficiency. It simplifies keeping hot things hot and cold things cold. There is a relationship between the temperature of a material and the reflection of light in the infrared portion of the spectrum. Energy can travel through space via radiation, and all bodies emit radiant energy across a continuous spectrum of wavelengths. Also called the emissive power, the energy flux is proportional to the fourth power of the absolute temperature of the emitter. When the wavelength is between 0.1 and 100 microns, the radiation is thermal. This radiation is the energy transferred by electromagnetic waves caused by the molecular motion of atoms that characterizes temperature.

A continuous spectrum of wavelengths makes up the thermal radiation emitted from an ideal material, called a black body, which absorbs all incident radiation. In reality, bodies are neither ideal absorbers nor emitters and behave as "gray bodies." The emissivity of the surface of a gray body is then the effectiveness with which radiation is emitted.

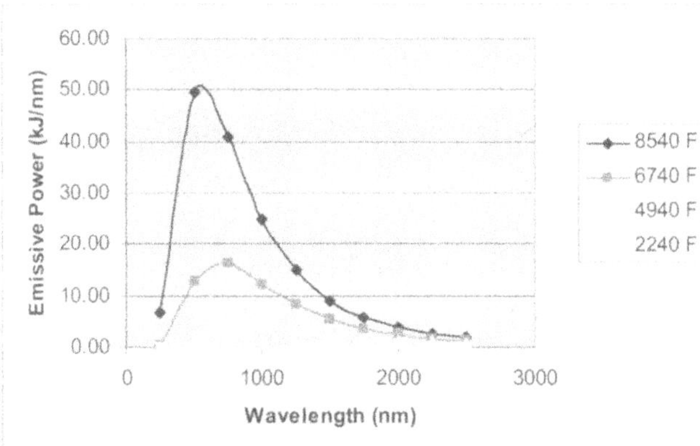

Figure 1. Spectral distribution of monochromatic emissive powers for an ideal radiator or black body at various temperatures

Figure 1 shows the spectral distribution of an ideal radiator over a range of temperatures. The amount of emitted radiation increases with temperature. The emitted radiation can then be absorbed, reflected, or transmitted by a second body such that:

$$\alpha + \rho + \tau = 1$$

where α = absorbtivity (the fraction of the heat flux absorbed by the surface), ρ = reflectivity (the fraction of the heat flux reflected), and τ = transmissivity (the fraction of the heat flux transmitted).[1] For opaque materials, $\tau = 0$, so

$$\alpha + \rho = 1$$

The phenomena of absorbtivity, reflectivity, and transmissivity are shown schematically in Figure 2. The two types of reflection are regular (or specular) or diffuse. When regular reflection occurs, the angle of incidence equals the angle of reflection. For diffuse radiation, the reflected radiation is scattered by the roughness of the surface and other factors. The value of each phenomenon varies with the material, its thickness, surface finish, and the wavelength of the incident radiation.

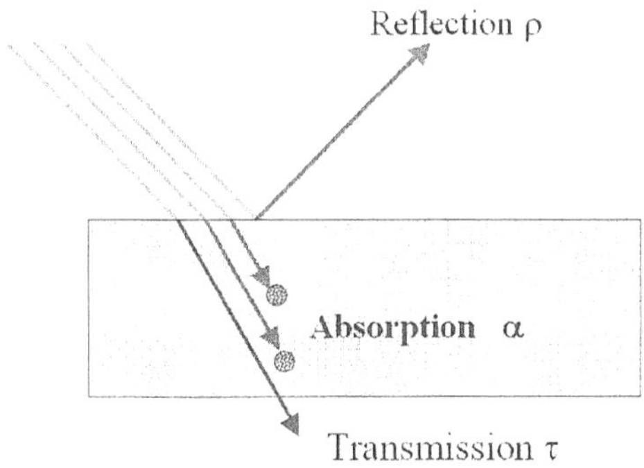

Figure 2. Behavior of incident light

The absorption, α, adds energy to the object through heat transfer. There are three modes of heat transfer across a temperature gradient:

1. Conduction – Through material with two surfaces at different temperatures
2. Convection – Heat transfer to and/or from a surrounding liquid or gas of a different temperature
3. Radiation – Transport of electromagnetic energy

Radiation emissive power is a function of the temperature of the radiation source, particularly in the infrared region. By increasing the infrared reflectivity of a material, the absorption and temperature increase can be reduced and the energy can be reflected back into space. First, an infrared-reflective finish can keep a surface cool. Second, an enclosure coated with an infrared-reflective can more efficiently heat the air inside. This was proven with a porcelain enamel cover coat specially formulated for maximum infrared reflectivity in a paper presented at the 2000 Porcelain Enamel Institute Technical Forum.[2] However, the coating described required a high thickness for maximum efficiency, and was inherently light in color.

Recent developments in materials science has resulted in the identification of pigments that have properties beyond the visible range of color, 400 nm to 700 nm. These materials, mixed metal oxides such as $(Fe, Cr)_2 O_3$ [black hematite], $(Al, Cr)_2 O_3$ [chrome alumina pink], and $(Al, Mn)_2 O_3$ [manganese alumina pink] represent materials with structures with substitution of various cations (Cr^{+3}, Fe^{+3}, Ti^{+4}, and combinations of these) that produce high infrared reflecting corundum-hematite structure pigments. For other metals in a rutile lattice (N^{+2} + $2Sb^{+5}$ for $3Ti^{+4}$ for example), coatings may be tailored to specific spectral responses. This pigments been used extensively for the EPA Cool Roofs initiative.[3]

To determine if a dark colored infrared-reflective porcelain enamel finish is possible, a clear frit was milled with dark infrared-reflective pigment and two different conventional black pigments. The mill formulation is shown in Table 1.

Raw Material	IR	Black 1	Black 2
Clear Cover Coat Frit	100	100	100
Clay	5	5	5
Borax	0.2	0.2	0.2
Bentonite	0.25	0.25	0.25
Magnesium Carbonate	0.2	0.2	0.2
IR Black Pigment	5		
PE Black Oxide 1		5	
PE Black Oxide 2			5
Water	42	42	42
Fineness	2-3/200 M	2-3/200 M	2-3/200 M
Fire	1520 F/3.5 min	1520 F/3.5 min	1520 F/3.5 min
Over White			
L	7.49	7.17	7.98
a	1.54	-0.45	-0.12
b	0.67	0.07	0.4
Over Ground Coat			
L	7.59	8.07	7.23
a	1.49	-0.41	0.06
b	0.59	-0.2	0.42

Table 1. Infrared-reflective enamel mill formulations

The enamels in Table 1 were applied either over fired white cover coat or ground coated steel. The reflectivity was measured with a Perkin-Elmer UV/Vis/NIR Lambda 19 spectrometer and is shown in Figure 3. Whether applied over white cover coat or ground coat, the enamel with the infrared-reflective pigment showed significantly higher infrared reflectivity. Therefore, the pigments developed for the Cool Roofs initiative could be used to tune heat reflectance of conventionally processed porcelain enamel. Possible applications envisioned include home cooking, fireplaces, pan supports, and other applications where the enamel is in close proximity to a radiant energy source for heating purposes.

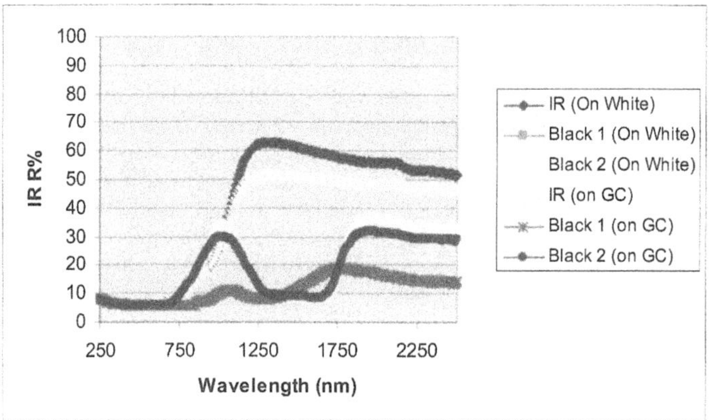

Figure 3. Infrared-reflectivity versus pigment type

Part 2 – Reducing Oven Energy Usage

The other improvement in the possibility of increasing energy efficiency with enamels evaluated is the replacement of the pyrolytic enamel used in oven cleaning with a coating using a much less energy intensive cleaning mechanism. Oven cleaning technology consists of three types: (1) self-cleaning pyrolytic ground coat, (2) non-self-cleaning ground coat, and (3) catalytic continuous clean enamels. The first reduces foodstuffs to ash with exposure to temperature between 900 and 1000°F (482 and 538°C), while the second requires harsh alkaline cleaners to remove soils. The third has largely fallen out of use and relies on high-metals, porous enamels to catalyze the reduction of soils to ash at normal cooking temperatures.

AquaRealEase® is a new alternative. It is porcelain enamel with a patented formulation that allows baked-on food residues to be released with exposure to water. As such, it offers a more environmentally-friendly option to cleaning ovens without the use of harsh chemicals, high temperatures, and the resulting fumes. Additionally, interlocks and extra insulation would not be required, freeing the oven manufacturer to add other features such as advanced electronics or integration into a home network to the previously old-fashioned range.

AquaRealEase® is applied with conventional processes. It has the mechanical durability and thermal resistance of traditional enamel, can be applied in a single fire to steel, and fires out between 1470 and 1570°F (799 and 854°C). Figure 4 shows a photograph of a commercially produced range with an AquaRealEase®-coated cavity.

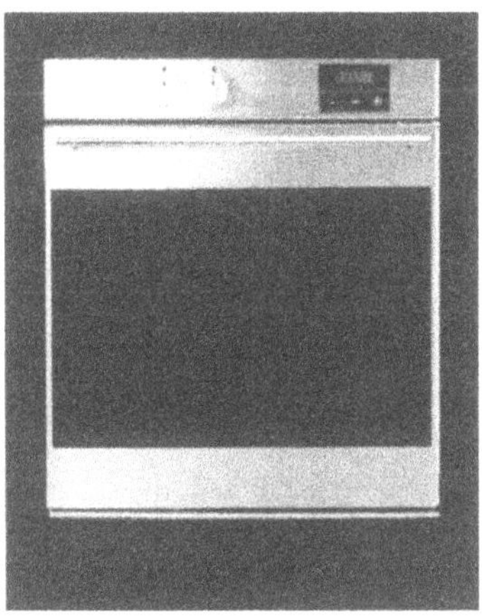

Figure 4. Commercially-produced AquaRealEase range[4]

A typical cleanability result for AquaRealEase® is shown in Figure 5. The cleanability has been assessed by the European FAN (Facile à Nettoyer, literally Easy-to-Clean) test. First, five steel rings are glued to the coating surface. Second, salted whole milk, gravy, lemon juice, egg yolk, and ketchup are placed in each ring. Third, the test panel was baked at 482°F (250°C) for 30 minutes. Then, the glue thermally decomposes, which allows the ring to be removed. After a brief immersion in hot water, all of the soils were easily removed from the AquaRealEase® plate but stains and residues were left on the conventional enamel plate.

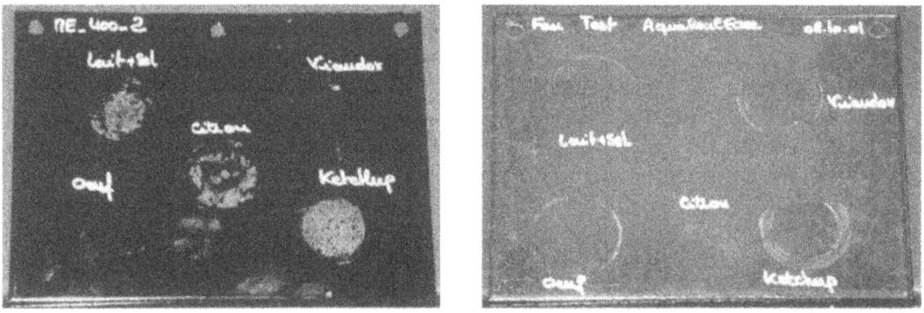

(a) (b)

Figure 5. Cleanability results for (a) conventional enamel and (b) AquaRealEase®

Although cleaning cycles may vary for various range manufacturers, AquaRealEase® sheds baked-on soils in service with exposure to water. One option is to generate steam with closed vents by using only the lower heating element to heat 1 L of water in a pan on the lower rack to 190°F (88°C) for at least 30 minutes. Then, the oven is turned off and allowed to cool for 30 minutes. The sides of the cavity are wiped with the soft side of the sponge to allow water to run down the sides. After 20 minutes, the soils can be wiped out of the range. Because of the use of steam for the cleaning cycle, AquaRealEase® has promise for steam cooking ranges emerging on the market.

AquaRealEase® is a durable coating based entirely on porcelain enamel technology that offers a low temperature cleaning cycle. Mechanical and thermal stresses from the pyrolytic self-clean cycle are eliminated. No harsh oven cleaners are needed. The finish is entirely inorganic, VOC-free, and has the hardness and scratch resistance of conventional porcelain. Additionally, energy usage and cost is reduced by eliminating the pyrolytic cycle.

In summary, two methods to use porcelain enamel to increase energy efficiency have been reviewed. First, infrared-reflective pigments can be added to enamels to reflect heat away from the enameled ware. Second, water-cleaning AquaRealEase® enamel creates the opportunity to eliminate the energy-intensive pyrolytic self-clean cycle from household ovens.

References

[1] D.R. Poirier and G.H. Geiger, *Transport Phenomena in Materials Processing* (TMS: Warrendale, PA, 1994), 370.

[2] Charles Baldwin, "Heat-Reflecting Enamel for New Ovens," *Proceedings of the Porcelain Enamel Technical Forum* 62, 2000.

[3] "Cool Roof Rating Council"
http://www.coolroofs.org/
(1 December 2009).

[4] "Porter and Charles"
http://www.porterandcharles.ca/ovens.html
(1 December 2009).

FROM DECORATIVE TO FUNCTIONAL PRINTING COLORS

C. Schlegel
PEMCO BRUGGE bvba

Presented by Peggy Damewood, Pemco Corporation

As both needs and technology have evolved, printing colors have changed as well. The following is a description of the evolution form decorative to functional color printing.

1. INTRODUCTION

The French word "décor" during the last few centuries, meant a means of artistic and promotional expression, a decoration. It has also represented an additional value for the decorated parts. A decoration, which is realized by printing, is not an exception to that rule and even if it remains closely linked to arts and designs, it is evolving more and more towards technical functions related to the arriving at new "high-tech" technologies. The ecological aspect is for its part not to be underestimated in that evolution; therefore:

- the flux is changing from lead-bearing to lead-free glass systems.
- the oil-based printing mediums are more and more replaced by water-based mediums.

PEMCO and particularly its Italian unit have been accompanying these continual changes for over 40 years by innovating in the fields of glass fluxes without lead, additives and printing mediums.

In order to recall the main points characterizing the printed "décor", you will find below most of the technical steps that have been incorporated in the printing pastes, their composition, preparation, application, curing (drying) and firing. The progressive changes that are supporting their evolution to more technical rather than decorative ends will also be presented.

2. WHAT CHARACTERIZED THE PRINTING PASTES OF YESTERDAY?

2.1. The composition

A printing paste generally includes three components:

- the glass flux
- the "additives" and
- the "medium"

2.1.1. The glass flux

Historicially glass fluxes were exclusively based on lead-bearing frits, while today allowing for the required thermal expansion, they are based on:

- "Bismuth silicate" – or "zinc borate" frits, if the substrate is glass.

- "Alumails" (V_2O_5-based) frits, if the substrate is aluminium or Al-alloy.
- "Alkaline borosilicate" frits, if the substrate is steel or cast iron.
- "Alkaline-earth, borosilicate" frits, if the substrate is ceramic.

In addition to its high fusibility and its ability to develop colors, the glass flux part is also influencing the gloss level and above all in the chemical durability. Glass fluxes may sometimes be pre-colored during melting.

2.1.2. The additives

Are primarily:

- Opacifiers, like TiO_2, SnO_2, CeO_2 or $ZrSiO_4$
- Pigments (stains)
- Some suspending agents, like colloidal materials

2.1.3. The medium

Which can be based on:

- "oils" (e.g. pine oil...)
- "glycols" (butyl glycol...)
- "waxes" (stearates...)

"Oil"- or "glycol"-based mediums are added in order to improve their printing and drying properties, generally modified in extending their properties using lubricating, fluidizing, thixotropy adding or cross-linking agents.

"Oil"-based mediums present 2 inconveniences:

- they have a distinctive pine oil odor
- they require the use of solvents to clean the printing equipment

"Glycol"-based (water miscible) mediums are maybe a little bit less robust to the printing process, but present the advantage to be easily cleaned using water.

"Wax"-based mediums, which are thermo-fusible and need heated printing means, which are more generally reserved for curved surfaces (e.g. beverage containers)

2.2. Preparation of the screening pastes

It is essential that the fineness of the solid component particles and in particular the glass fluxes are easily and completely wetted by the medium. It becomes clear that the preparation and more particularly the grinding of these components is very important and desirable to incorporate a classification step. Dry grinding is preferred; it also limits the glass flux solubility. The water- or alcohol-based wet grinding continues to be used, but has declined in popularity due to performance and safety issues.

The preparation of the printing pastes, also called printing inks is generally comprised of three steps:

- the preparation of the various components based on their quantities and specifications
- the mixture of the solid components with the printing medium in order to realize a homogeneous paste
- the refining of the paste using a "3-roll mill" refining step in order to eliminate the coarse particles or agglomerates and to improve the paste behavior on the printing screen.

In case a correction is necessary, liquids are simply added by mixing; solids on the contrary, in order to be homogeneously dispersed, need to be refined as previously defined. It also needs to be underlined that some precautions are necessary concerning the storage of printing inks and more particularly concerning those prepared with UV curing mediums.

2.3. The application techniques

Regardless of the substrate type, they need to be clean and free of grease and dust.
In order to allow bond, stainless steel and aluminium substrates generally need to be previously grit blasted or brushed before printing in order to increase the surface areas.
The application techniques used have not evolved a lot concerning their method but much more concerning their automation. Those techniques are:

- flat or rotative screening
- screening with indirect "décor" transfer or pad printing
- roller coating

All those techniques, except for the one by roller coating which is based on a transfer by engraved cylinder, are using screens as transfer means.
Those screens which were yesterday woven using:

- natural silk threads,

are today manufactured using synthetic threads, like:

- nylon mono-threads, more robust and presenting a higher elasticity
- polyester or polyamide multi-threads, different through their very high chemical resistance.

Sometimes those screens are manufactured using

- metal threads: in order to be usable as heating elements to fluidize thermo-fusible pastes.

2.4. The curing (drying) and firing steps

In order to make printed parts easier to handle, the curing (drying) of the printing inks needs to be thoroughly realized. Drying is an important factor in the process speed of the manufacturing lines. Therefore, it affects productivity.

Drying can be implemented using:

- hot air
- IR-radiation
- UV-radiation

Contrary to organic printing inks that only need to be cured, mineral printing inks also need to be fired after drying (curing). This firing step is generally a function of the substrate to be printed, for instance:

- between 540 and 680°C for aluminium and glass
- between 750 and 850°C for steel and cast iron
- between 880 and 1080°C for stoneware ceramics
- between 1050 and 1200°C for stoneware or porcelain

The furnaces used for this purpose are often muffled on the one hand to guarantee a neutral atmosphere and on the other hand to avoid the corrosion of the gas heated radiant tubes. Electrically heated elements are generally avoided, as being too expensive.

3. WHAT IS CHARACTERIZING THE MINERAL PRINTING INKS TODAY OR WILL LIKELY CHARACTERIZE THEM TOMORROW?

- They will continue to be based on 3 families of components but the latter will all have been reviewed in order to satisfy new needs and specifications.
- Those new needs or specifications, for a big part, result from the explosion of the electronic, tele-visual, hi-fi, internet and automobile industries, at a lower volume from the appliance industries and without doubt at a very low level from the conventional decorations applied over glass, steel or ceramic substrates.
- Among those new functions, we will in particular find dielectric, half-conductive, conductive or inductive functions. Additionally to those functions, the ink layers need to be bubble-free and present the lowest diffusion grade possible in order not to affect their performances.
- Application-wise, those needs generally imply the use of screens proposing much more elaborated weavings, like the use of threads with cross sections different from cylindrical ones or the use of coated threads in order to speed the passing of the inks or to avoid the effects of static electricity that can negatively influence the production rate.
- Those requirements of more functionality in order to produce layers with additional or improved performance to increase productivity or to realize new decorative effects, can only be achieved through the following milestones.

3.1. Changes in the printing paste formula itself

3.1.1. Concerning the flux part

The fluxes (frits) without PbO and CdO should also be free of chemical elements that are presenting a too high mobility, such as alkalis and boron in order to minimize the diffusion rate in the substrate. On the other hand they should also possess a finer

particle size in order to achieve photographic resolutions. They should additionally have the highest possible density, as near as possible to those presented by lead-bearing frits, so that a high filler grade can be achieved, limiting organic residues if the firing cycle should be completed under worst conditions.
Bi_2O_3-based frits, whose developments have accelerated forward despite their high cost due to their meeting the previously mentioned specifications; Bi_2O_3-based frits (fluxes):

- present a high density
- are particularly fusible
- are chemically resistant
- are transparent and able to match a maximum of colors, including yellow and red cadmium colors and
- should be able to be mixed with high quantities of additives without any quality decrease (bubbles, pinholes etc.)

Today our activities in the field partly consist of limiting the alkalis in order to avoid their diffusion in case of multi-layer printing systems and partly of reducing Bi_2O_3 in order to be more market competitive.

3.1.2. Concerning the additives

The main additives are generally:

- the mineral-based pigments
- the opacifiers, such as TiO_2, SnO_2 and CeO_2

PEMCO is proposing a full palette of mixable pigments in order to be able to match most target colors. The opacifiers are used in addition to the pigments in order to match "pastel" colors. However, as printing pastes are more and more evolving from coloring to more technical functions, the opacifiers are either evolving in their quality or are replaced by additives that are introducing other properties, like dielectricity, semi-conductivity and conductivity, magneticity, tribology, anti-staticity, luminescence or "bubble structure" regulation.

Among these additives, you can find:

- the conventional opacifiers TiO_2, SnO_2 or CeO_2 but nanosized.
- the zirconates and titanates
- colloidal silica or alumina
- metals: Ag, Pd,...etc
- etc...

The choice of the additives and their particle size will be achieved according to the searched function, which may often be much more technical than decorative and will require highly precise photographic resolutions.
During that evolution, where colors will play only a secondary role, the pigments will lose most of their interest and will often only serve to list the various layers necessary to build the searched functionality, as for instance it would be the case in a printed electrical system. They should therefore not be underestimated, as some of them are presenting interesting magnetic properties.

In the automotive industry and particularly concerning the so-called "black paste" framing automotive glasses, we can show a notable evolution of their functionality due to productive issues; in fact the black frame, which appears on automotive glass, and yesterday was just a protection against UV light, should next propose an anti-stick, recrystallizing phase in order to allow jointly the hot processing deformation of 2 glass sheets; the black frame will stick on the glass sheet, on which it has been printed but not stick on the recovering glass sheet, which has been deformed during the same hot processing step. Such an evolution has in a short time become a reality.

3.1.3. Concerning the mediums

The printing mediums are without doubt the links in the printing chain that have experienced the most important evolution.
In fact in order to fulfill all requirements, from environmental or applicative origin, linked to drying or firing steps, the mediums go through notable changes.
The oil-based mediums are replaced by water-based mediums in order to eliminate the use of solvents to clean the printing screens; concerning this point, please don't forget that the "emulsion" used to create the screen image needs to be chosen as a compliment to the chosen medium. The mediums that have to fulfill new functions each day, are becoming more and more complex in their formulas, which often appears to be disturbing due to their combustion during the firing cycle; in fact the organic residues that are remaining trapped in the layer, are a source of inhomogeneous bubble structures yielding sometimes bubbles with a diameter near the layer thickness, of surfaces that are losing gloss grade through the numerous "pinholes" present or of poor color development; those residues can also sometimes be advantageous, particularly when dark colours need to be intensified. Even though infrared drying mediums are mostly used today, ultraviolet drying will for production performance reasons, without doubt become the preferred medium in the future.

3.2. Printing paste preparation

Printing paste preparation has not changed a lot; the particle size of the different components remains the main parameter to guarantee the best quality. However, we should not forget that sub-micron flux particles are easily going into solution and are greatly influencing the molten viscosity. Those sub-micron particles can also activate the photoinitiators present in the UV mediums, which can originate from a premature polymerization and consequently agglomerate.

3.3. Evolution concerning the application techniques

The most used techniques are flat or rotative printing using silk screens (flat or cylindrical). Innovation activities were particularly intense in those fields. In order to facilitate the screening output, to compensate viscosity differences of the printing inks or to ensure design sharpness, the screen quality, texture and weaving have been adapted. The thickness of the emulsion layer necessary for screen manufacturing being today easily adjustable, it becomes possible to deposit in one pass high paste thicknesses; however "the higher the thickness, the worse the design resolution.

Next to conventional silk screen printing, other "decorative" application technologies are also growing using mineral printing inks, whose particle sizes are often close to or finer than the pigment fineness (< 1 micron); "inkjet", "rotocolor" or "transfer laser" are among those technologies. The following are without doubt the most promising:

- "Inkjet" printing is particularly performing well when organic inks are used, a little bit less when abrasive mineral inks are used. The "inkjet" printing is transferring the color using a measuring nib presenting an opening diameter less than 1 micron, which easily explains the required particle size and the problems generated by the ink abrasiveness (problems that can be limited by increasing the lubricity grade of the inks).
- "Rotocolor" printing, borrowed from the tiles industry, is based on a transfer by roller, which has been engraved by laser technology allowing on the one hand a high precision and sharpness and on the other hand the décor transfer without any substrate stop.
- "Laser" transfer is a technology borrowed from laser printing on paper. This technology uses solid inks, also called mineral toners. However, it remains an expensive technology.

In case of "inkjet" or "laser" transfer techniques, all decorations are first displayed on a screen, so that it will always be possible to revisit and optimize them for color or aspect-wise.

3.4. Concerning the drying and firing steps

The drying and firing steps are without doubt the steps that know the lowest evolution. Even if drying by UV exposure results in increased productivity, it doesn't progress as fast as hoped and consequently IR drying remains the most popular drying technique. The firing steps are also not quickly evolving and continue to depend mostly on radiation. Even though microwave firing seems to be an interesting alternative, it has rarely been introduced since it is considered too expensive energy-wise.
Similarly "laser firing", which in one step transfers the picture and completes its firing, being complex in its implementation and limited color-wise, doesn't really, except for marking (logos), highly succeed on the market.

4. CONCLUSION

The serigraphic printing technologies are no longer as they used to be, only meant for decoration of ceramic or glass objects, such as tiles, china, packaging glass, architectural glass or even vitroceramic heating plates...etc. The techniques are more and more used for technical purposes, where the function is becoming more important than the decoration or color. This escape to new fields of use accompanied by a multitude of innovations more or less important, mostly originating from specific needs, is accelerating day by day, as the printing techniques, through their high degree of automation and flexibility and the use of more and more sophisticated products, enable strong productivity improvements.

ISO9001:2000 – WHY DO IT?

Kara Joyce Kopplin
QTEC Consulting

This presentation will review the benefits of ISO 9001:2000 compliance and the financial gains realized by companies with strong quality systems.

WHAT ARE THE GOALS OF ISO9001:2000?

Defect Prevention
Customer Satisfaction
Continual Improvement
International Recognition

These are achieved through the discipline of the quality standard, which serves to strengthen corporations by building controls into all of the business processes. When processes are properly developed and controlled, they yield good products and good customer service. By analyzing performance and taking appropriate actions, the business will continually improve.

HOW WILL WE ACHIEVE THOSE GOALS?

An effective quality system will control the manufacturing and business processes to:

- REDUCE Variation – INCREASE Consistency
- REDUCE Scrap and Rework
- MINIMIZE Waste
- INCREASE Efficiency
- INCREASE Customer Confidence and Satisfaction
- OPEN THE DOOR for new business
- CONTINUALLY IMPROVE

The simple model for control and continual improvement is

Plan ------- Do ---------- Check ----------Act ---------- Repeat!

The entire ISO9001:2000 standard is modeled after this continual improvement approach, and is represented as follows:

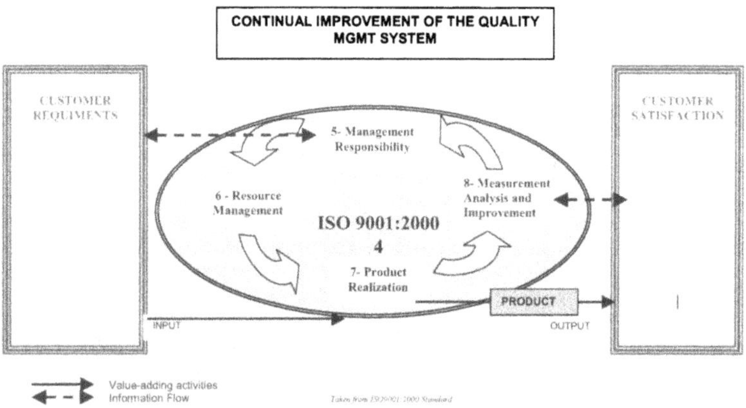

WHAT IS THE GLOBAL COMPLIANCE TREND?

Compliance is no longer the exception, it's the rule in the global economy.

- The ISO Survey – 2006 shows a **16%** rise in certifications to ISO 9001 in 2006...**900,000** certs in **170 countries** by the end of 2006...up from 775,000 in 161 the previous year

- ISO/TS16949...and ISO 13485 increased by more than **60%** ...28,000 certs in **78 countries**, up from 5,000 in 67 countries the previous year

ASQ *Quality Progress*, February, 2008

HOW DO COMPANIES WITH EFFECTIVE QUALITY SYSTEMS PERFORM FINANCIALLY?

They enjoy higher profits than the companies that don't follow a quality system philosophy.

- "Q-100 Index is a subset of the S&P 500...using companies with the highest relative composite quality scores."
- "Since inception in 1998, Q-100 gains were at more than 41% at the end of 2004, S&P was only 30%".

ASQ *Quality Progress* April 2004

WHY WERE THEY MORE PROFITABLE?

Without a disciplined quality system, inconsistency, waste, inefficiency and lack of control can eat away at your business, not just as scrap, but in DOZENS costly of ways.

The cost of poor quality are extensive, and expensive, and fall under the general categories of costs of prevention, appraisal, internal failures, and external failures.

(list with permission from Accuride Wheels, Henderson, KY)

Prevention
- Quality Training
- Process Planning
- Design Review
- Quality Engineering
- Preventive Maintenance
- Setup Scrap
- Receiving Inspection
- Supplier Quality Evaluation

- Appraisal
    - Inspection
    - Calibration
    - Testing
    - Quality Audits
    - Inspection Equipment
    - Inventory Evaluations

- Internal Failure
    - Production Scrap
    - Supplier Related Failure
    - Repair
    - Rework
    - Sorting
    - Failure Analysis
    - Corrective Action

External Failure
- Returned parts
- Warranty parts
- Shipping
- Travel
- Appraisal/Inspection
- Failure Analysis
- Corrective Action
- Process Changes
- Design Changes
- Scrap
- Allowances
- Liability Costs

HOW DOES ISO9001:2000 COMPLIANCE MINIMIZE THESE COSTS?

With the discipline of defined processes to control the entire business, the cost of poor quality noted above can be minimized or eliminated. The specific requirements of the quality standard include tools and disciplines that support this approach, including

- design reviews
- design verification and validation
- supplier control
- standardized work instructions and production control
- equipment and gauge PMs and calibration
- employee training
- data analysis and management review

WHAT HAVE WE CONCLUDED?

By implementing ISO9001:2000, your company will realize the value-added, cost-effective benefits of

- Rework / Scrap Minimization
- Increased Customer Confidence and Satisfaction
- Continual Improvement
- International Recognition

NEW EU CHEMICAL REGULATIONS WILL SIGNIFICANTLY IMPACT USA AND GLOBAL BUSINESSES: *REACH*

Jack Waggener, PE
URS Corporation

"REACH"
(Registration, Evaluation, & Authorization of Chemicals)

- New "EU" Chemical Regulation
- June 1, 2007 (Initiation)
  Next 10 years: Phased Approval
- Pre-Registration: June to December 2008
- Impacts imports to EU
  (and exports from EU)
- Goal: "REACH be used globally"

REACH = Registration, Evaluation and Authorization [and Restriction] of Chemicals

- Came into force on June 1, 2007
- An EU "regulation", meaning it is immediately applicable and needs no transposition into national legislation
- Gradual introduction of the obligations through to 2018 (if pre-registered)
- Puts all responsibility of demonstrating the safety of chemical substances on industry, not government. Reversing burden of proof
- Replaces most of the current EU directives and national laws on chemicals

Pre-Registration of Existing Substances

- You can avoid some supply chain risk by pre-registering substances in their products – *even if they suspect that registration will not be necessary*.

- Pre-registration allows for substances to continue to be used until the full registration is completed - benefit from a potential 11 year phase-in of registration

- The pre-registration period is June 1 – December 1, 2008.

If a substance is not pre-registered, after December 1, 2008 it cannot be manufactured/imported into the EU until it has been registered!

REACH Timeline

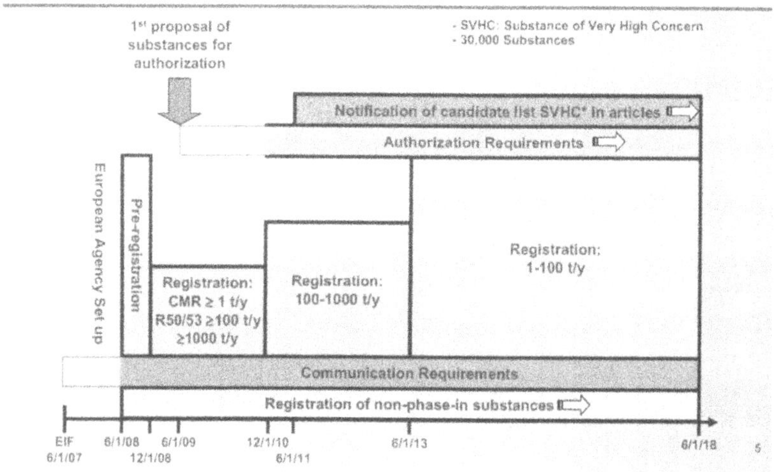

Company Impact

- Economic Impact – What is at risk?
- Competitive advantage for EU companies with or within the EU or if they sell to a company who operates in the EU?
- Affect on the supply chain?
  - Reduction of suppliers
  - Increased supplier management costs
- Lost business opportunity & market access in Europe?
- Potential impact to existing deliveries?

And inevitably…

- Some substances may:
  - not be on the market

(*suppliers will choose not to register*)

  - be in limited supply
  - be more expensive
  - be phased out

- Some will be withdrawn due to 'authorization'

- The chemical industry is a global marketplace, so the effects will be global

REACH Applies to the EU
(All Imports & Exports)

- 27 Member States of the European Union are:

Austria
Belgium
Bulgaria
Cyprus
Czech Republic
Denmark
Estonia
Finland
France
Germany
Greece
Hungary
Ireland
Italy
Latvia
Lithuania
Luxembourg
Malta
The Netherlands
Poland
Portugal
Romania
Slovakia
Slovenia
Spain
Sweden
United Kingdom

- Norway, Iceland and Liechtenstein
  - considered WITHIN the EU for REACH (these countries will also apply the regulation)
- Switzerland is OUTSIDE the EU

REACH Requires?

- The same level of risk assessment and analysis for existing chemicals as required for registration of new chemicals

- The reduction in risk for the 30,000 most frequently used substances throughout their life cycle

- The gradual withdrawal of substances of very high concern from the EU market

Implementation and Monitoring:
European Chemicals Agency based in Helsinki, Finland
REACH will lead to the withdrawal of substances from the EU

REACH Regulates….
Substance - a chemical element and its compounds in the natural state or obtained by any manufacturing process

Examples:

- Methanal (Formaldehyde) – CAS No 50-00-0
- Nickel metal – CAS No 7440-02-0
- Tetrachloroethylene (Perchlorethylene) – CAS N° 127-18-4

Phase in substance - a substance having an EINECS number (European Inventory of Existing Commercial Chemical Substances)

Preparation - *mixture* or *solution* composed of two or more *substances*

Examples: paints, glue, lubricants, varnishes, adhesives, alloys, P.E.

Intermediate - *substance* that is manufactured for and consumed in or used for chemical processing in order to be transformed into another substance [*non-isolated; on site isolated; transported isolated*]

Polymer –*substance* consisting of molecules characterised by the sequence of one or more types of monomer unit. Such molecules must be distributed over a range of molecular weights. Source: ECHA Guidance for monomers and polymers, March 2008

- Article - object which during production is given a specific shape, surface or design which determines its function to a greater degree than its chemical composition (*Important definition*)

The shape, surface or design is more important than the substances it is made of.

What are Articles?

- The product that you make is probably an article

Appliances	Control Gear
Engine	Gearbox…
Assembly	Electronics

- The spare parts you make are articles (component, spares-kit, maintenance kit, test equipment)

BUT, be careful, they may contain regulated substances (e.g. adhesives)

- You might BUY articles – close-to-form forgings, bolts, memory chip…

ALL Substances are to be Known and Assessed...
All substances… on their own;

or contained in preparations (mixture of at least two chemical substances *without* the formation of new substances) or contained in articles
(All "ingredients", raw materials, consumables, components, alloys and their constituents)

"Only Representative"

- To comply with the registration requirements of REACH, you will need an "Only Representative"
  - "Only Representative" - EU-based natural or legal person acting on behalf of a Non-EU manufacturer fulfilling the duties of an importer
  - As of April 2008, the "Only Representative" will need to file a separate registration for each substance/legal entity it represents
- Non-EU companies can Register on their customer's behalf, but only by using an "Only Representative"

Estimated Registration cost for one Substance

Volume tier	Individual registration (euro)[1]	Average testing cost cost substance (euro)[2]
1 - 10 tonnes	1,600	56,360
10 - 100 tonnes	4,300	279,838
100 - 1000 tonnes	11,500	799,562
> 1000 tonnes	31,000	1,582,616

[1]Draft Regulation on Fees and Charges, Individual Submission under Article 6, 7, or 11, October 2007, p. 16.
[2] Source: Fleischer, M. "Testing Costs Testing Capacity According to REACH Requirements - Results of a Survey of Independent and Corporate GLP Laboratories in the EU and Switzerland," September 2007, pgs. 96-114

Some EU manufacturers/suppliers may look at these costs and decide that the cost of REACH compliance is not worth "it" and terminate some products

Implementing *REACH* – What Have We Got to Do?

- REACH requires a cross-functional effort, involving but not limited to:
  - Purchasing
  - Contracts
  - Engineering and Design
  - Manufacturing
  - Legal
  - ESH……

- REACH will require new information to be collected within companies, and to be passed up and down supply chains

Conclusions

- Many Corporations Taking Action
- Make Sure Your Company Is!
    - Use REACH To Your Advantage
    - Do Not Be Left Behind!
- Other Issues:
    - REACH in China, Canada, Etc.
    - U.S. Legislation
    - Rohs (Pb, Cr(+6), Hg, Cd)
    - Etc.

A HEAT RESISTANT LABEL MATERIAL AND ITS APPLICATION IN HIGH TEMPERATURE PROCESSING

Huimin Yang, PhD
Brady Corporation

This paper addresses the problems of making labels heat resistant at temperatures of 700 and 900 degrees Celsius on both ceramic and steel surfaces.

New Product Objective
To provide a low cost, pre-ceramic label material for identifying both ceramic and metal parts that go through a high temperature process (>700°C).

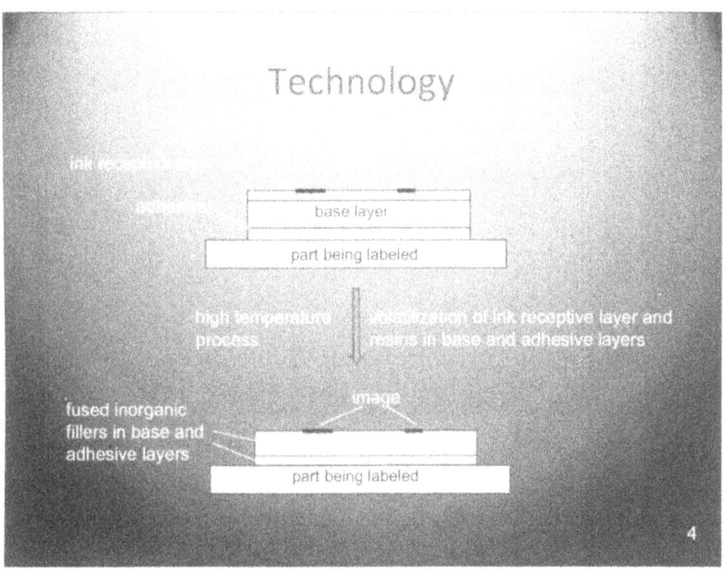

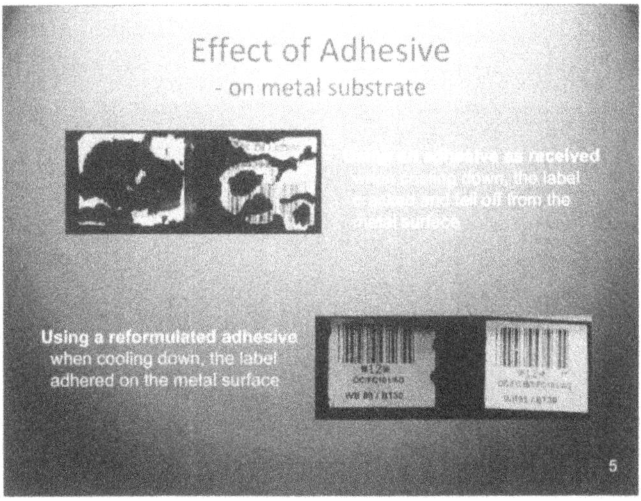

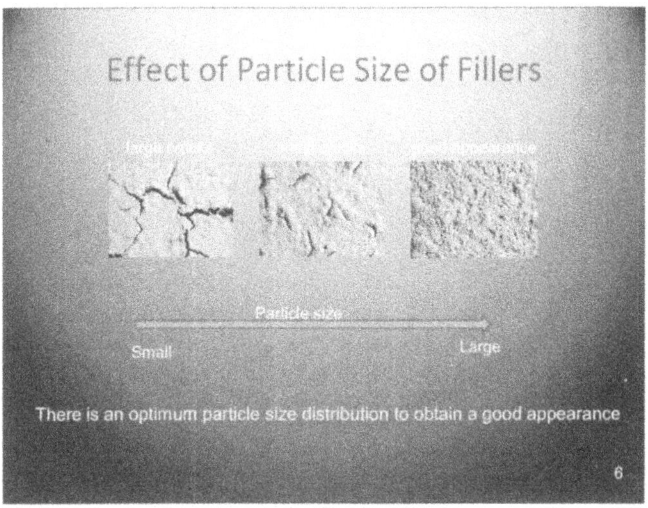

CONCLUSION

A low cost, pre-ceramic label material was prepared for identifying an object that need to go through a high temperature process. By carefully formulating each layer of the label, pre-ceramic labels can be applied for identification on both metal and ceramic surface.

FURNACE MOISTURE ISSUES

Jason Butz – Coating Technician
Engineered Storage Products Company
DeKalb, IL

INTRODUCTION

Furnace moisture plays a crucial role in the firing of porcelain enamel coatings. For this reason continuous monitoring of furnace moisture during the firing process can help to reduce moisture related defects.

This paper will be presenting a general overview of some furnace moisture defects, and their causes, as well as ways to monitor furnace moisture. It will also present data gathered at Engineered Storage Products Company (ESPC) that shows how seasonal humidity levels effects furnace moisture and ways to help reduce defects caused by furnace moisture.

TYPES OF FURNACE MOISTURE DEFECTS

Although some of these defects can be caused by more than just moisture in the furnace, ESPC has found that they tend to occur when the percent moisture volume *(%MV)* in the furnace is higher than 2%. Some of the most common defects found are:

- Black Specks
- Holidays
- Crawling
- Orange Peel
- Blistering

As the ambient temperature starts to increase so does the %MV in the furnace. Therefore these defects tend to rise during the summer months. This leads to the next section the causes of higher furnace moisture.

CAUSES OF FURNACE MOISTURE

Moisture in the furnace can be attributed to effects such as products of combustion, mill additions, moisture in the coating, and a plants relative humidity [1]. Although there are other causes these are typically the most common.

Products of combustion can be a result of poorly maintained or damaged radiant tubes, or exhaust systems that allow natural gas to leak into the furnace chamber [2].

Once the coating is in bisque form, *mill additions* such as clays and bentonite have the potential to hold in moisture. This moisture is then released while the ware is going through the hot zone during the firing process.

Moisture in the coating is also commonly caused by the improper drying of the ware. When the ware has not been fully dried it enters the furnace and releases the excess moisture into the furnace atmosphere.

Moisture in the coating can be also attributed to the *relative humidity in the plant* which is one of the main causes of improper drying. This is why moisture related defects tend to be higher during the summer months, when relative humidity is higher

DETECTING FURNACE MOISTURE

There are a couple of ways in which moisture in the furnace can be monitored. Peter Vodak lists three ways in his paper "Continuous Measurement of Furnace Moisture" [3]. They are as follows:

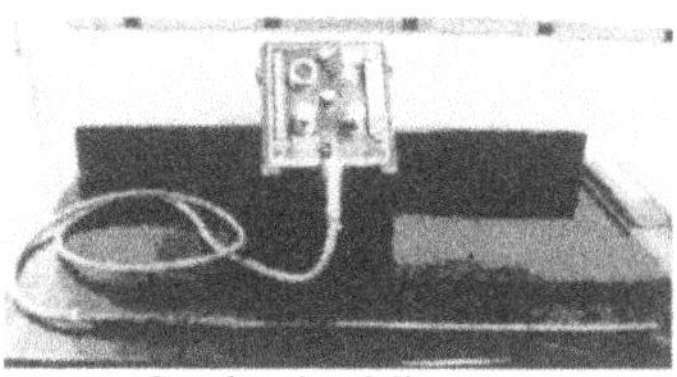

Condensing Mirror

Dräger Tube

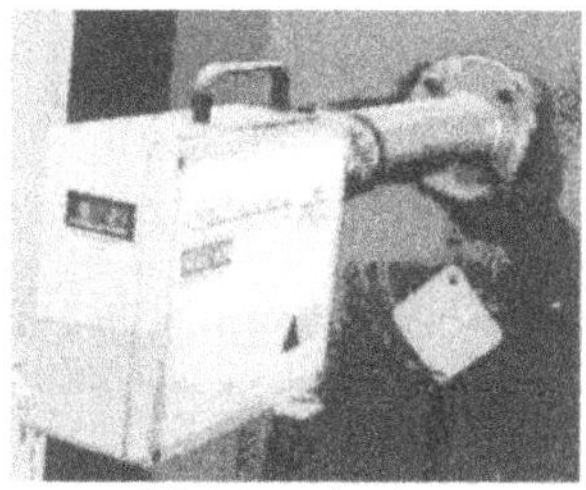

Moisture Analyzer

Each of these ways works well to monitor moisture in the furnace, however the high temperature continuous moisture analyzer is the only one that can continually monitor the moisture and is not susceptible to human error. ESPC is currently using the MAC 120 moisture analyzer to continually measure the %MV in their furnace and have been using it since 2003. It has helped to lower the amount of moisture related defects. This is because it allows ESPC to react quicker to decrease the %MV in the furnace.

ESPC SEASONAL MOISTURE RELATED FINDINGS

Earlier In the paper seasonal moisture changes were discussed as an issue in dealing with furnace moisture. Since 2007 ESPC has been monitoring its plants daily ambient temperature, dew point, relative humidity and the furnace percent moisture volume to help in them in understanding seasonal moisture changes and there effects on defect rates. In Fig.1 below you can see that as the temperature rises, so does everything else including the % MV inside the furnace (Close up of %MV in Fig. 2).

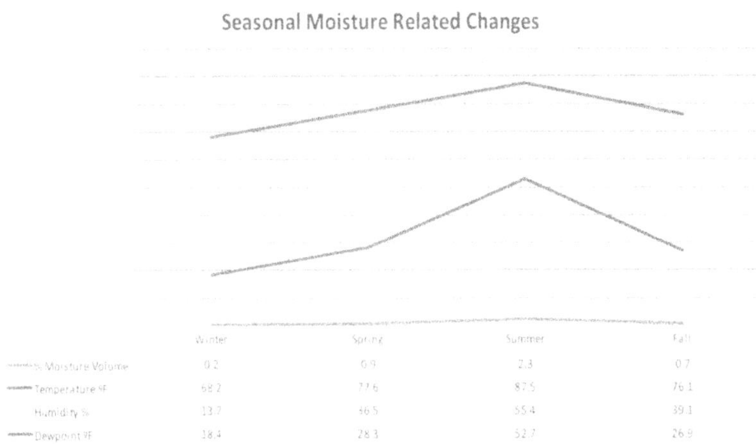

Fig. 1

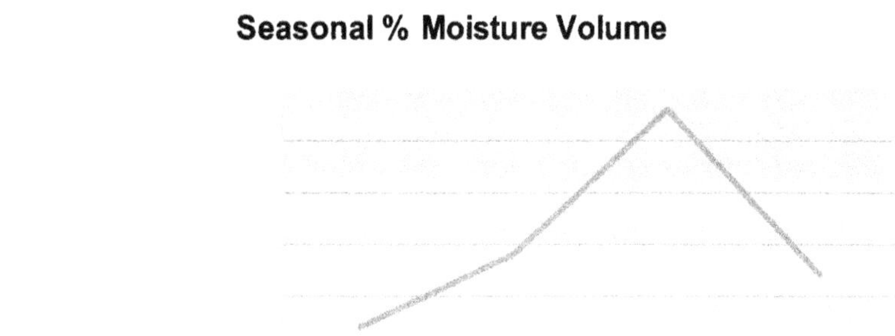

	Winter	Spring	Summer	Fall
% Moisture Volume	0.2	0.9	2.3	0.7

Fig. 2

ESPC noticed that as these entire readings rose, so too did the percent of moisture related defects. There was almost a 50% increase in moisture related defects from the winter months of December through February to the summer months of June through August. This amount of defect rate increase has caused ESPC to come up with ways to decrease the amount of moisture related defects that occur.

WAYS TO REDUCE MOISTURE RELATED DEFECTS

There are several ways in which moisture related defects can be reduced. For example, a preventative maintenance walk through of the furnace to check that all radiant tubes and exhaust systems are working properly should be performed regularly. This will reduce the chance that products of combustion will be entering the furnace atmosphere and causing an increase in furnace moisture. ESPC performs such checks once every two weeks, with more frequent checks if a problem is suspected.

Another way to reduce moisture in the furnace is by injecting clean dry air into the furnace. This method works to reduce the amount of moisture in the furnace because it absorbs the moisture and also forces it and any products of combustion out the furnace vents. ESPC currently uses this method by having a regenerative desiccant air dryer hooked up to it furnace. However the dryer is turned on manually and if the furnace operator forgets to turn it on during the furnace start up on a humid day then there is a greater chance for moisture related defects to occur. For this reason ESPC is currently looking to connect the Mac 120 moisture analyzer directly to the air injection dryer so that when the moisture in the furnace gets too high it will inject dry air into the furnace. In this way the furnace operator does not have to remember to turn the dryer in.

Another possible solution to moisture related defects is proper drying of the ware before it enters the furnace. This can be accomplished by making sure that adequate dryers are set up. This insures that no moisture will get trapped in the coating before it enters the furnace. ESPC is currently looking into additional drying to make sure that this is no longer a problem during summer months when the relative humidity is higher since that seems to be the time when proper drying is an issue.

CONCLUSIONS

From the information that has been gathered at Engineered Storage Products Company one can see that a plant's environment can have a large effect on moisture related defects. This is all the more reason to continuously monitor furnace moisture to try to prevent the defects that it can cause. With the proper preventative maintenance on the furnace and improved drying techniques along with the moisture monitoring equipment defect rates can decrease.

REFERENCES

1. R Ott. "Effect of Furnace Moisture on Ground-Coat Surface Quality"; Proceedings of the PEI, Volume 44 (1982), pp.370-373.
2. H. Evele. "Effects of Furnace Moisture on Enamel Quality and Recommended Levels"; Proceedings of the PEI, Volume 67 (2005), pp.95-104.
3. P. Vodak. "Continuous Measurement of Furnace Moisture"; Proceedings of PEI, Volume 67 (2005), pp105-112.

FURNACES AND FIRING

Mike Horton
KMI Systems

The furnace is the key process in the porcelain enameling process which affects both the part quality and the part costs. This review of the technology and explores ways to maximize the efficiency of the heat treatment step in the process.

In today's economy, the demands on the porcelain enamel furnace are greater than ever before.

With energy and alloy price increases, the firing process of the furnace needs to be re-evaluated to optimize the porcelain enamel operation.

Within the past two years the cost of alloy used in burner tubes and manifolds has more than doubled.

Natural gas and electricity costs are rising.

With the current economic changes in most operations, the enameling process must also change to meet the new demands.

With these process changes there maybe a reduction in maximizing the furnace operations. A furnace audit needs to be scheduled on a regular basis.

A proper maintenance schedule of the furnace can reduce the operation cost.

The purpose of this paper is to review the furnace design, operation, process and maintenance items to help reduce the operation cost.

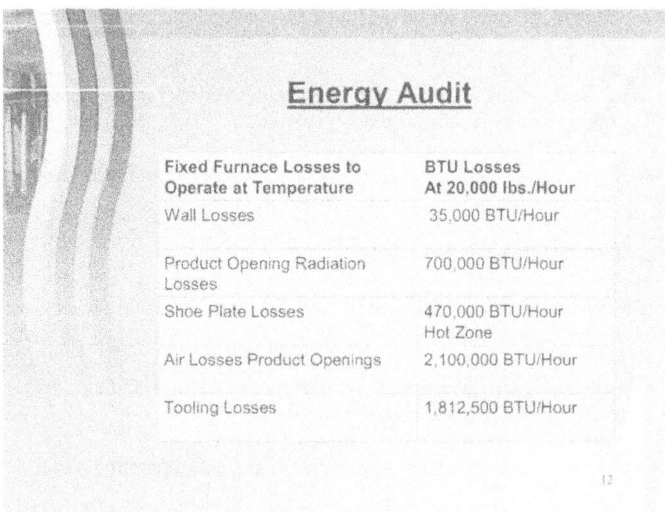

Energy Audit

Fixed Furnace Losses to Operate at Temperature	BTU Losses At 20,000 lbs./Hour
Wall Losses	35,000 BTU/Hour
Product Opening Radiation Losses	700,000 BTU/Hour
Shoe Plate Losses	470,000 BTU/Hour Hot Zone
Air Losses Product Openings	2,100,000 BTU/Hour
Tooling Losses	1,812,500 BTU/Hour

12

Operation Furnaces Losses
Exhaust Losses: 1,475,500 BTU/Hour

Part or Ware Losses
(Beneficial Losses):
1,875,000 BTU/Hour

Total BTU/Hour 8,468,000

Air Seal Operation

All continuous operating furnaces, regardless of wet or powder coating firing, need two air seals at the entrance and the exit to minimize the air losses.

For example: A furnace entrance which has a measured outgoing air stream of 400° F at 350 ft/min with a 4 ft x 7 ft high opening will have a heat loss of :

Air Loss

Air Loss is 350 ft/min x 4 ft wide
x 3 ft high flow area =
4,200 acfm actual cubic feet per minute

Standard Air Loss

$$4{,}200 \text{ acfm} \times \frac{460\text{ F} + 70\text{ F}}{460\text{ F} + 400\text{ F}}$$
=
2588 scfm Losses

Heat Losses to Air Infiltration
(One Opening)

2,588 scfm x 1.08 BTU/Hr F scfm x 1300 F = 3,633,550 BTU/Hour
At $10.00/1,000,000 BTU/1,000
Cu Ft Natural gas x 3,633,550
BTU/Hour =$36.34/Hour X
5,000 Hours/Yr = $181,700/Yr

Tooling to Ware Ratio
Weight of Tooling : Weight of Parts Fired
With the current production demand changes, the tooling and the ware mix must be re-evaluated to optimize the loading and to reduce the energy consumption. New lighter weight tooling should be examined for reduce energy.
Tooling to Ware Ratio
Desired Ratio 40% or Lower

Tooling Design
Tooling should be evaluated for current part design and weight.

Lighter weight tooling may be tested for part support, maximum loading density, and life cycling.

Furnace Maintenance for Operation Cost Reduction

Furnace Shell and Ceramic Fiber Insulation
Inspect the ceramic insulation on the wall and roof for damage and leakage.

Repair any roof damage where parts hit the fiber and shoe plate slot and seals.

Replaced damaged bricks and compressed ceramic fiber.

Air Seal Inspection
Check air seal for damper settings, any wrecks or distortion of the air vanes.

Check the intake return duct for blockage with powder frit, parts and hooks.

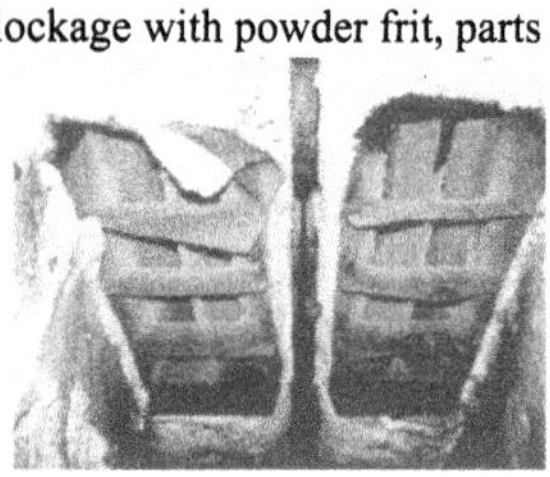

REPAIR DAMAGE AIR SEALS
Clean the duct system.

Check with a calibrated air flow meter the cold air velocities and record.

Check the fan belts, fan wheel and the housing and its insulation cover.

Product Opening Radiation End Losses
Repair ceramic fiber baffle walls to reduce radiation losses.

Where possible, add additional baffle walls.

Replace any damaged center walls.

Furnace Tuning and Maintenance
All burners need to be tuned at least once a year to reduce energy and burner tube life.

Once a month, the alloy expansion joints should be inspected.

All fans, exhaust duct, dampers and duct insulation must be inspected monthly.

Combustion air filters must be cleaned and inspected.

At operation temperatures, a visual inspection through view ports should be made of the burner tubes and manifolds.

The conveyor shoe plate wear, fit over the furnace track, expansion joints and the lubrication should be inspected.

MAKING PORCELAIN ENAMEL MORE COMPETITIVE WITH STAINLESS STEEL: A HOLISTIC APPROACH TO PROCESS IMPROVEMENT

Ken Kaluzny
Coral Chemical Company

A holistic approach that considers the entire enameling process is discussed in light of making porcelain enamel a more competitive alternative to stainless steel for appliances.

Master of the Obvious

- Increasing Manufacturing Costs
  - Raw material & chemical costs
  - Energy & labor costs
- Compete With Asian Manufacturing
  - Low cost, Chinese labor
  - Government subsidies
- Porcelain Enamel
  - Market share lost to stainless steel fabrication
  - Need to bring market back to P/E

Asian Manufacturing

- How does this relate to the presentation?
  - Increased stainless steel demand
  - Limited stainless steel supply
  - Produces higher stainless steel cost
  - Increased stainless steel cost leads to higher consumer prices
  - Higher consumer prices is driving manufacturing to China/Asia
- Personal Mantra

Porcelain Enamel Finishers

- North American Porcelain Enamellers need…
  - Compete with stainless steel fabricated appliances
  - Compete with Asian produced products
- Quality manufacturing is essential
  - Integrity of finish
  - Integrity of product's function
  - Environmental and safety issues, ie. Lead in paint
- Holistic Approach to Manufacturing
  - Minimize operating costs
  - Maximize profitability

Porcelain Enamel Institute

- As an organization...
  - We need to bring back to P/E market share
  - Establish new market trends
- We all need to work together to keep manufacturing domestic and porcelain enamel competitive.

Stainless Steel

- Stainless Steel Alloys
  - 150 grades
  - 15 common commercially
  - Variety of finishes
- Increased Stainless Steel Demand
  - Relatively fixed supply
  - Leads to getting "close" matches
  - Change in alloy changes forming characteristics
    - Increased nickel content produces forming blow-outs
    - Decreased nickel content produces orange peel

Stain & Corrosion Resistance

- Stain Resistance
  - Is it truly **stainless** steel?
  - Water spotting issues
  - Thermal induced staining
- Corrosion Resistance
  - Variable corrosion resistant alloys
  - Non-passivated stainless steel
  - Poorly passivated stainless steel
- Production Quality Issues

Fabrication Issues

- Compared to steel, stainless steel...

    - is more difficult to form
    - requires greater lubricity
    - produces greater scrap cost
- Stainless steel peculiarities
    - Stainless steel finish integrity
    - Broadcasting slug marks
- Lubricity Implications
    - Lubricants are harder to remove
    - Lubricants are more damaging to cleaner baths

Post Forming Issues

- Cleaner Implications
    - Increased cleaner consumption and cost
    - Purified water requirements
    - Alcohol wipe after processing as stainless shows all blemishes
- Forming issues likely to produce scrap
    - Scratches can't be buffed out
    - Imbedded fines

Stainless Steel vs. Porcelain Enamel

- Fabrication
    - Advantage – Porcelain Enamel
    - Chemical Related Costs
    - Higher scrap costs with stainless steel
- Non-conforming Fabrication
    - Enameling Steel is easier to rework pre-fire
    - Grinding imperfections from stainless
    - Post-fire argument
- Porcelain is more stain resistant
    - Heat and food resistance
    - Water spotting
- Porcelain is more durable
    - Corrosion resistance – *Glass doesn't rust*
- Color Options
    - Stainless steel has less finish options
    - Porcelain Enamel has greater color options
    - Decaling

The Holistic Approach

Definition

- **Requires Team Work**
    - Team members consisting of departments
    - Fabrication
    - Pretreatment
    - Vendors
    - Purchasing
    - Environmental Health and Safety

Control is the "Name of the Game"

- Certify Raw Material Quality
    - Insuring metal certification
    - Lubricant and pretreatment certificates
    - Insuring coating/frit certifications
- System and Solution Control
    - Maintain operating parameters
    - Maintain systems
- Environmental
    - Health and Safety
    - Waste Disposal

The Holistic Approach will focus on…

- Reducing manufacturing costs
- Maximizing line speed
- Maximizing line density
- Minimizing energy consumption
- Minimizing chemical operating costs

The Holistic Approach:
Chemical Systems

- Selecting right lubricants
- Consider other fabrication operations
    - Piercing and beading
    - Welding
    - Redrawing
- Environmental Health and Safety
    - Eliminate oil from process
    - Reduce regulated substances

The Holistic Approach: Automation

- Lubricant applicator
    - Too much lubricant
        - blow outs and wrinkling
        - extra chemical cost
        - greater cleaner cost
    - Too little lubricant
        - produces tears
        - Scrap costs
        - Increased die maintenance
- Cleaner Controllers
- Automated quality inspection

The Holistic Approach: Mechanical Systems

- Non-Chemical Related Systems
    - Optimizing the P/E coating & firing process
    - Single chain system
    - Line density and speed

The Holistic Approach: Implementation

- Brainstorming
  - What issues need to be addressed?
  - What restricts productivity?
- Everything is running Fine
  - Forming Lubricants
  - Low Temperature Cleaners
- Manufacturing Paradigm
  - Don't fall into the trap

Case Study 1

- Appliance Manufacturer
  - Eliminate defect on oven collar
  - Cost savings ~$135,000 through scrap reduction
  - Radius steel/Flow at radius/Altered conductivity
  - Changed lubricant so that it was weld-able
  - Eliminate hand wiping this radius with volatile solvent (acetone)
  - Had some fires with solvent

Case Study 2

- Use of chlorinated paraffin based lubricant
- Caused flash rust within 24 hours
  - Orange rust at 36 hours
  - Hygroscopic lube
  - Chloride liberated from heat of draw
  - Lubricant change provided 3 weeks rust protection
    - Improved cleaning/parts easier to clean
    - Increase line speed
    - Increase productivity

Case Study 3

- Grill Manufacturer
  - Extended bath life 4/6 weeks to 52 weeks
  - Over 85% bath life increase
  - Labor reduction
  - Increased line speed 2 fpm (10-12%)
  - Cleaning related rejects dropped/disappeared
  - Eliminated roughly 10 recharges
  - Improved die life
  - Eliminate spot weld defects
  - Sludge reduction

Commentary

- Porcelain Enamellers can become more efficient
  - Stainless steel fabrication issues
  - Manufacturing team work
  - Departmental rivalries
- We need to invest in our future
  - Manufacturing Paradigm
  - Think long term
  - Innovation

- Develop New Strategies
  - Generate Market Pull
  - Piggy-back Advertising
  - Power of the Internet
- New Market and Options
  - Using virtues of Porcelain Enamel
  - Color and Designer Options
  - Decaling Untapped

REQUIREMENTS FOR THE USE OF PORCELAIN ENAMEL IN INDUSTRIAL COATINGS

Hemant Dandekar
Porcelain Industries Inc.

A review of the requirements that porcelain enamellers need to achieve in order to compete effectively with other coatings and materials in today's changing industry.

Outline

- Size of process opportunity
- Types of heat exchangers
- Key issues in heat exchangers
- Requirements for coatings
- Potential path forward

Size of Process Opportunity

- US industrial sector consumes 32.5 Quads or 32.5x1015 BTUs per year
- Petroleum, petrochemicals, iron and steel, food and pharma represent 12.4 Quads
- Onsite losses in the top 5 industries are 4.22 Quads ($46 Bn/year)

Nature of Onsite Losses

- Poor insulation,
- Inefficient heat transfer
- Lack of recovery of exit streams such as flue gases and contaminated steam.

Potential for Immediate Impact

- Waste heat recovery in gas and liquid streams for chemicals, refining and forest products - $2.2 Bn/year
- Improved process heat transfer for chemical and refinery applications $860 million/year

Process Equipment

- Heat Exchangers
- Distillation Columns
- Reactors
- Storage Vessels
- Pumps/compressors

Types of Heat Exchangers

- Tubular
  - Shell & Tube
  - Double Pipe
  - Coiled Tube
- Applications
  - High temp. & pressure
  - Gas to gas & vapor condensation
  - Power generation, process industries, chemicals
  - Versatile, large scale

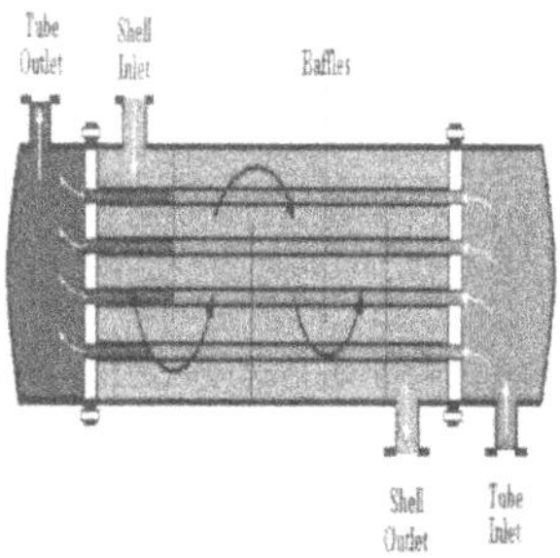

Plate Heat Exchangers

- Type
  - Gasketed
  - Spiral
  - Plate coil
  - Lamella
- Use
  - Liquid-liquid applications
  - Dairy, pharma, chemicals, off shore, forest products
  - Compact design, scalable, easy to maintain

Extended Surface Exchangers

- Type
  - Tube finned - liquid to gas
  - Plate Finned - gas to gas
- Use
  - Low Ht Tr. Coeff. Gas-Gas or Gas-liquid
  - Finned surface to increase heat transfer surface
  - Use in cryogenics, air cooling/heating
  - Land based water/oil coiling heat exchangers

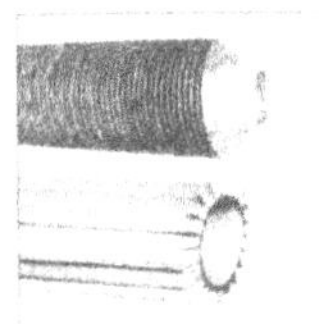

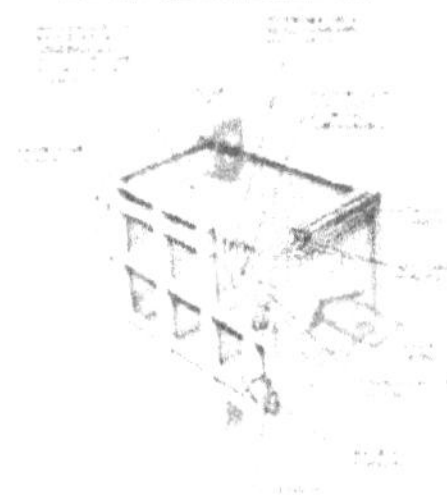

Key Operating Parameters

- Temperature range
  - Plate exchangers 100-350 oF
  - Tube heat exchanger up to 1200 oF
  - Fin tube - -100 oF to 700 oF
- Chemical composition of fluids
  - Acidic composition of flue gas in power refinery applications (H_2S, SO_2)
  - High temperature H_2 attack of substrates
  - Acidic composition of food components
  - Salt water resistance for sea water applications
- Presence of slurry, suspension and pulps
  - In forestry applications, slurries can require high abrasion resistance

Opportunities

- Replace expensive alloys by coated/enameled carbon steel
  - Lower material cost
  - Improve operating costs
    - Reduced fouling
    - Increase periods when tubes/plates are replaced
    - Reduce/eliminate frequent assembly/dissassembly
- Enable waste heat recovery at a lower capital investment
- Isolate metal surfaces that catalyzed by product formation in heat exchangers

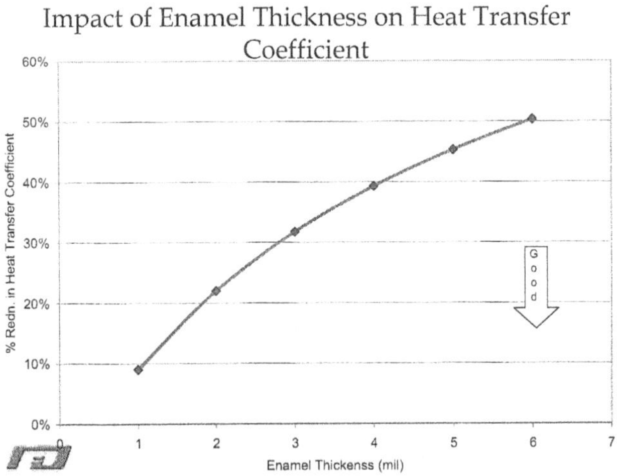

Coating Requirements

- Flexible – ability to coat thin substrates such as heat exchanger plates of thickness 0.5 to 1.2 mm, channel gaps 1.5 to 5 mm and dimensions from 0.03 to 2.2 m.
- Thin coating – avoid insulation penalty.
- Ability to coat complex geometries
  - Low viscosity/fine particle size in slip
  - Fine particle size in powder

Wet enamel process (dip, flow or spray)

- Pros
  - Highly customizable formulation
  - Low capital cost
  - Can customize surface finish
- Cons
  - Difficulty controlling thickness below 3 mil
  - Inability to coat complex parts
  - Flexibility of coating (due to thickness)

Powder process

- Pros
  - Can control thickness
  - Ability to coat edges and complex geometries
- Cons
  - Limited formulation variety
  - Not able to customize surface
  - May not coat interior sharp corners

Sol-gel coatings

- Pros
  - Can be uniformly coated to very fine thickness- less than 1 mil
  - Surface can be functionalized with hydrophobic characteristics
  - Thin coatings allow flexibility and ability to coat complex shapes
- Cons
  - Unproven technology not scaled yet
  - What are the maximum temperatures it can take?
  - Might require multiple applications to achieve desired thickness of coating

D. Raybould et al. US Patent 6,670,050 (2003)

- Sol-gel coating for Titanium heat exchanger
- Zirconia and Alumina based solgels
- 3 or more dip steps
- Firing temperature 300 to 500 oC
- Prevent thermal oxidation of Titanium heat exchangers up to 750 oC

D. Barrow et al. US Patent 5,585,136 (1996)

- Substrate: Stainless steel, mild steel, Inconel
- Zirconia Propoxide sol-gel modified with Ca, Y and Sr cations and hybridized with fine ceramic powders of Zirconia and Alumina
- 20 or more repeated steps with spin coating
- Firing temperatures of 700-1000 oC

Brinker & Reed US Patent 4,476,156 (1984).

- Partially hydrolyzed sol-gel solution with dispersed fine glass particles
- The sol-gel forms a chemical bond with the substrate
- The film is then "cured" at 50 oC above the glass softening temperature ~450 oC to densify the film
- Single step process with thickness up to 100 microns (4 mil)

Conclusions

- Significant opportunities in the process industries to improve heat transfer and reclaim lost heat and improve yield
- Conventional enamel coatings can be limited in their use due to thickness and flexibility
- Patent literature shows promise with hybrid sol-gel with ceramic/glass powders on a variety of substrates
- Industry needs further research/development of novel approaches to create next generation of coatings

REACTIVE VITREOUS ENAMEL COATINGS FOR CONCRETE REINFORCING STEEL

Cullen Hackler
Porcelain Enamel Institute

The significance of achieving a strong bond between protectively coated reinforcing steel and concrete in construction of bridges, highways and buildings.

Fundamental Concepts

- Porcelain enamel designed to adhere to steel
- Fusion of glass to steel at high (+/- 1500°F)
- Reactive ingredient dispersed in and adhered to the glass coating – portland cement
- Reactive component hydrates and integrates into the concrete as the structure is poured and cured

The Problem

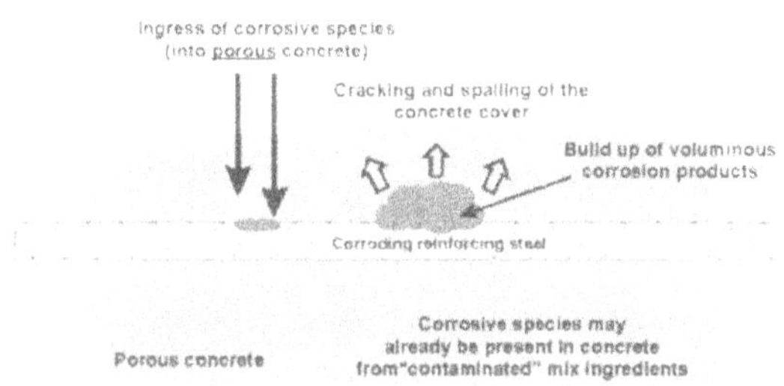

What have we accomplished?

- Glass coated steel rod
- Reactive ingredient on or near the surface
- Coating is fired at 1500° F
- Samples imbedded in concrete for pull-out testing

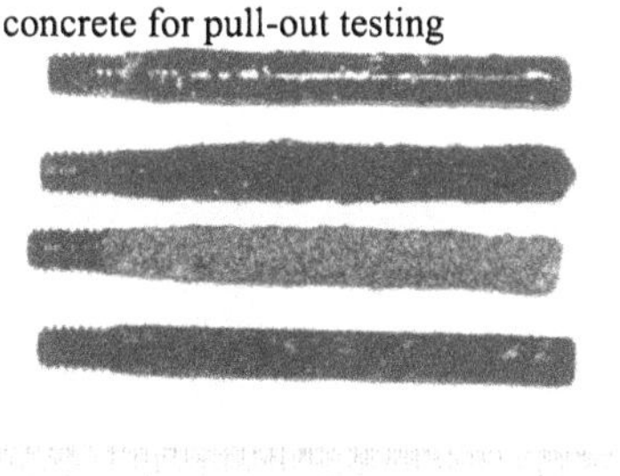

Structural Implications

- Reinforcing steel bonds to concrete
    - Rebar and welded wire mesh
    - Coated steel fiber
    - Corrugated and steel sheet
    - Wall ties and other building components
- Development length is reduced
    - Cost and weight savings
- End products exceed standards
    - Higher strength
    - Better performance in earthquake zones
    - Less cracking thus less water penetration

Potential Markets and Sizes

	Steel (lb)	% of market	Sq. Ft.	Enamel (lb/ sq. ft.)	Total frit pounds
Rebar (1" bar)	6,000,000,000	10%	58,883,333	0.146	8,596,967
Steel fiber	20,000,000	50%	39,255,556	0.146	5,731,311
Welded wire mesh	500,000,000	15%	29,441,667	0.146	4,298,483
Anchors, ties & joints	250,000,000	20%	19,627,778	0.146	2,865,656
Corrugated steel decks	400,000,000	15%	23,553,333	0.146	3,438,787
Architectural sheet	50,000,000	10%	1,962,778	0.146	286,566
Flooring	100,000,000	10%	3,925,556	0.146	573,131
Total					25,790,900

Coated Rebar Cost Comparisons

Material	[illegible]	[illegible]	[illegible]
Black iron (uncoated)	$0.32	$0.00	$0.00
Porcelain enamel-coated	$0.345	$0.025	$0.25
Epoxy coated	$0.42	$0.10	$1.00
Hot-dipped galvanized	$0.47	$0.15	$1.50
Epoxy II	$0.47	$0.15	$1.50
Purple Epoxy	$0.56	$0.24	$2.40

Plans and activities for 2008 and beyond

- Coated rebar
  - CERL and U S Army
  - MO DOT and Roesch
  - Ballistic Technologies
  - University of Louisville, UK, KY DOT and S & S Porcelain
- Coated steel fiber
  - ERDC blast panel study
- These will all provide additional data for us to use to revise and strengthen the "value proposition"
- Develop an ASTM test specification for P/E reinforcing steel types
- Optimize the porcelain enamel system(s) – water heater, alkali ground coat
- Understand the role of the portland cement and its "location" in the coating – surface layer, throughout, etc.
- Prepare samples and prototypes for testing and evaluation in construction products/projects and Execute commercial license with U S Army – summer 2008
- Continue to compile data and understand potential markets.

Value Proposition - Draft

- The glass coating actually protects the steel from corrosion and allows direct bonding to concrete (ASTM B 117, 2000 + hours and ASTM C 876, no corrosion)
- With the 5X increase in the bond strength of the reinforcing steel to the concrete - savings in both reinforcing steel and bulk concrete
- Many products and/or markets, where concrete is reinforced with steel, can benefit from this technology; rebar, fiber, welded wire mesh, corrugated, sub-flooring, building components, etc.

Conferences and Trade Shows

▪ NACE "Corrosion 2008"	March 16 – 20, 2008
▪ ACI	March 30 – April 3 and Nov. 2 – 6, 2008
▪ AASHTO	May 4 – 8, 2008
▪ PTI "Concrete Bridge Conf."	May 4 – 6, 2008
▪ ASHE	June 11 – 15, 2008
▪ NPCA	Sept. 18 -22, 2008 & Feb. 20 – 22, 2009
▪ PCI "National Bridge Conf."	Oct. 5 – 8, 2008
▪ PCA "World of Concrete"	Feb. 3 – 6, 2009

2009 PEI Tech Forum

ENERGY COSTS IN WASHER SYSTEMS

Brian Rozdilsky
KMI System Inc

The presentation will review the specific topic of energy consumption in washer systems and discuss strategies to conserve.

Areas for improvement in energy costs in washer systems are:

1. Process Temp
2. Insulate Tanks
3. Adding High Eff Motors
4. Install VFD's on Pumps
5. Review Washer Design
6. Counter Flow Stages

Where does the energy go? It depends on the required process temperature. Dropping 10 °F from 160°F to 150°F (71°C to 66°C) saves 550,000 BTU/hr (583 MJ/hr). Dropping 10 °F from 150°F to 140°F (66°C to 60°C) saves 400,000 BTU/hr (422 MJ/hr). Dropping 10 °F from 140°F to 130°F (60°C to 54°C) saves 290,000 BTU/hr (306 MJ/hr).

Insulating tanks can save 26,000 BTU/hr (27.4 MJ/hr).

Upgrading to high efficiency motors saves 140 KW/Year.

Adding variable frequency drive (VFD) will reduce energy usage by 8,614 KW per motor.

If the wash system is oversized, reducing the mask opening can save 100,000 BTU/hr (105.5 MJ/hr). The wash system should be sized for speed and for the part length.

Translated into cost savings, this is:

Lower Process Temp 10°F:	$8,000 to $14,000 per year per stage
Insulate Tanks:	$ 561 per year per stage
High Efficiency Motors:	Saves $125 per year per motor
Install VFD's on Pumps:	$ 603 per year per pump
Stream Line Washer Design	$2,000 to $ 13,000 per year

For example, assuming:

2 Shifts
$10/MBTU
$.07 KW/HR
5 Stages -15 HP
2 Heated Stages @160 F
240 Days/year

The savings are:

Lower Process Temp 10°F in Two Stages:	$ 30,720.00
Insulate Tanks:	$ 2,204.00
High Efficiency Motors:	$ 260.00
Install VFD's on Pumps:	$ 6,025.00
Stream Line Washer Design:	$ 36,376.00

TOTAL SAVINGS = $75,585

21ST CENTURY CLEANING SYSTEMS

Ken Kaluzny
Product Manager
Coral Chemical Company

This paper will discuss advances in cleaner chemistries and their advantages in current P/E process lines.

Suppliers of cleaning systems for manufacturing coating processes in the twenty-first century are facing a number of challenges. Many are living in a survival mode. Global competition has been affecting business decisions for some time now. Innovation is essential now more than ever. We need to be innovative with not only our products, but with our processes as well. Successful American manufacturing operations are meeting the challenges of outsourcing and fierce overseas competition via innovation, investment, lean manufacturing, and training. Innovative products, finished goods or manufacturing chemistry, help prevent us from competing solely on price. Process innovation is essential to our ability to compete and maintain a profit.

Environmental issues have driven change in the United States as well as in the porcelain enamel industry. Whether legitimately created by man or not, global warming is a hot topic. It appears that global warming is a sustainable term to the point of begetting new terminology. Marketers were far behind and "green" is a term that has been great for marketing. However, becoming "green" can help cut operating costs too. Simply stated, the 3-R's of "reduce, reuse, and recycle" can be applied to porcelain enamel pretreatments. Whether a trendy marketing approach or not, applying "green" attributes to cleaning operations creates twenty-first century cleaning systems.

Ambient cleaners are becoming more common but are not really new development. In fact, ambient cleaners have been around for a long time. In the twenty-first century, or particularly the current business climate, ambient cleaners are getting renewed attention as energy costs are high and sales are typically down. The use of the word ambient is somewhat ambiguous. Ambient in North Dakota is going to be different than ambient in Florida in most any month of the year. Low temperature cleaning would be more accurate and appropriate terminology for the pretreatment process. Some soils just need some heat to become fluid, making the cleaning chemistry more effective.

Other continuing trends are driven by environmental health and safety. The elimination of "bad actors" continues at a definite, yet uncertain pace. This could be the result of waste water discharge regulations or from corporate policy to reduce or eliminate SARA Title III reporting. The elimination of bad actors is also driven by the cost associated with disposal. Phosphates are commonly used in alkaline cleaners and are high on the bad actor list in states such as Minnesota. The trend is continuing to other states as well. Neutral cleaners are also a recent trend that is driven by environmental health and safety issues. They can work, but they aren't for every cleaning process.

Figure 1 contains a graph that shows energy savings resulting from lowering temperature incrementally 10 degrees. The absolute numbers relate to a particular set of energy costs, tank size and operating time. For a better idea of potential savings, ask the cleaner vendor. The relationship of temperature and cost is not linear. In other words, dropping cleaner temperature 10°F produces more savings at 150° F (66°C) than it does at 120°F (50°C). As the operating temperature decreases, the savings potential from temperature reduction drops. Running at 80°F (27°C) sounds like a huge energy savings if operating at 140°F (60°C). It is a big savings, but if reject parts start being produced, the energy savings will be offset and possibly eliminated. For every system there is a sweet spot for the cleaner temperature in which it will remove soils from your metal without producing rejected ware.

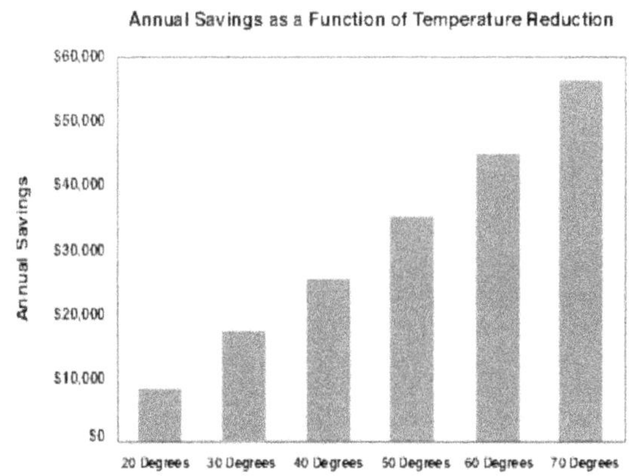

Figure 1. Energy savings with ambient cleaning

All of the 3-R's can be applied to water. Counter-flowing rinses can reduce water consumption while maintaining quality. Closed loop systems involve multiple rinses incorporating counter-flow rinsing and reverse osmosis filtration in the secondary rinse, which also reduces chemical usage as well. Running at low or ambient temperature will also reduce water usage as evaporative loss is minimized. However, in the case of a close loop system, if ran at low temperatures, water does have to be discharged at some point.

Counter-flow or closed loop systems will reduce chemical usage. Low temperature products don't necessarily reduce chemical usage. Process changes need to be verified and validated. Rinse aids can be used to reduce energy costs. Rinse aids reduce surface tension making water drain and subsequently dry quicker. This affords the ability to turn dry-off ovens down and reduce energy consumption.

There are other ways to be innovative with your manufacturing process. The right selection of metalworking fluids and cleaners can reduce cleaning costs as much as 40%. Choosing the right lubricant can increase your first pass efficiency by reducing rejects, reducing cleaner impact, and reducing cleaner usage.

Spray nozzles can even reduce energy consumption. As nozzles wear, they break up the solution into more droplets, increasing the surface area for heat transfer. The energy loss from "atomization" is at least as big as the energy required to heat the tank to operating temperature. Taking this a step further, the type of nozzle and operating pressure can either cost or save energy.

In conclusion, ambient cleaning is not new. However the current business climate is driven by economical and ecological factors. Applying green principles can lower your operating costs to help you survive and thrive in today's business climate.

USING CONTROL SYSTEMS TO GO "GREEN" WHILE SAVING YOU MONEY!

John O'Connor
ECC Controls of Illinois

This paper will detail how electric, gas, and water usage can be reduced in the finishing area while increasing quality and production. Detailed examples of systems that have focused on going "GREEN" will be discussed, including actual cost saving documentation. The necessary tools to audit finishing systems to see how they can go "GREEN" to reduce costs will be described.

Process controls in the porcelain finishing area can reduce chemical, electric, water, and gas consumption in washers and ovens. They can also improve production and quality rates with respect to conveyors and parts tracking.

There are several opportunities in process control areas. First, in the pretreatment washer, chemical usage can be reduced by maintaining proper chemical concentration, temperature and tank levels. Electric, gas, and water reductions can be achieved through washer pump, burner control and level controls. Counter flow rinses can be used to reduce water consumption as well. Second, conveyor control and monitoring provide line speed control, downtime reports, and monitoring of unscheduled stoppages, which can be synchronized with the oven to reduce gas usage. Third, dry off and cure oven controls allow adjustable set points as well as scheduling and stand by modes to reduce gas. Finally, parts tracking shows where parts are to activate energy saving control strategies and gives production rates information.

Pretreatment washer chemical controls maintain accurate concentrations and charge up using AccuFlo pumps. Automatic additions are made based on manual titration entries with several control modes can be used for adding chemicals and controlling rinse contamination. An example is shown in Figure 1. Mechanical variables in pretreatment washers include spray pressure control by variable frequency drive (VFD), temperatures, tank levels, nozzle blockage, and filter condition.

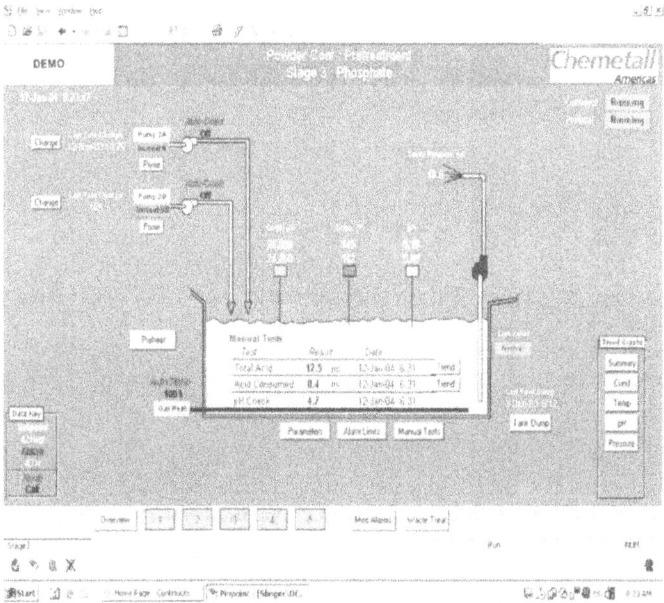

Figure 1. Pretreatment washer controls

Figure 2 shows an example of conveyor tracking. This documents line stoppages and speed, counts parts, and creates downtime reports.

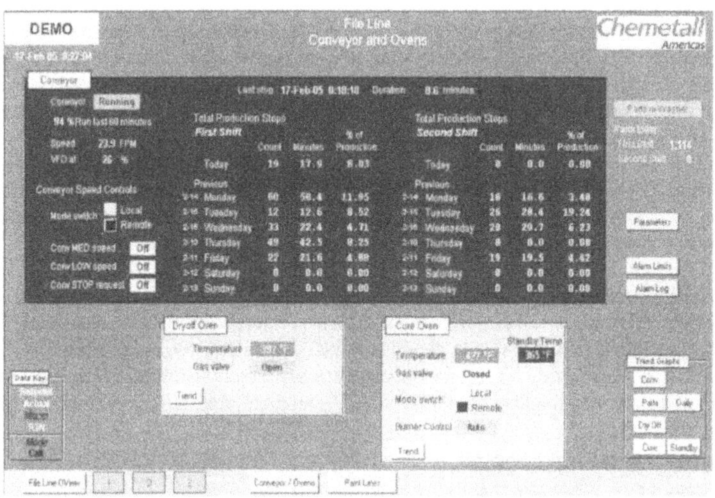

Figure 2. Conveyor control

Figure 3 shows oven controls. These include burner controls, stand-by-mode, automatic start-up and/or shut-down, and alarm temperatures by zone.

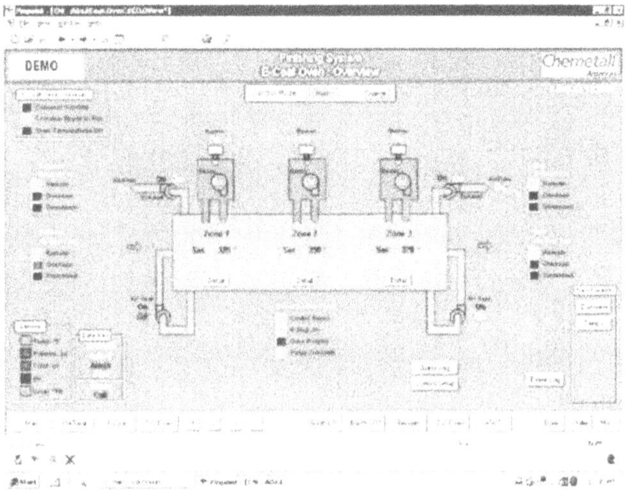

Figure 3. Oven controls

Finally, parts tracking controls are shown Figure 4. These provide data on the location and quantity of parts, the number processed, and washer conditions when parts were processed, as well as cycling ovens off if no parts are detected.

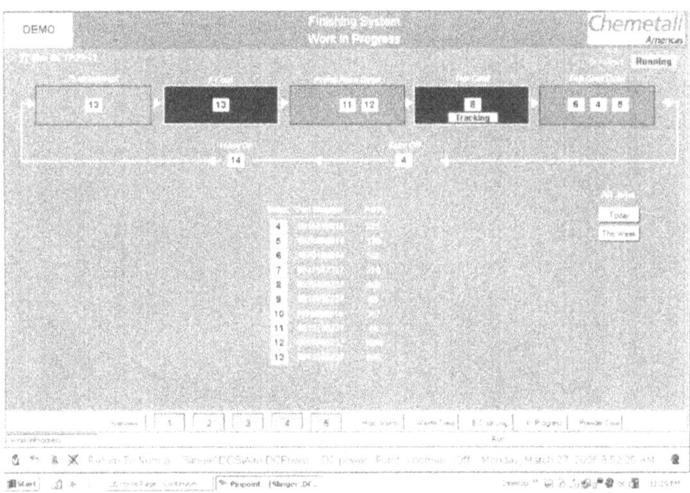

Figure 4. Parts tracking

PORCELAIN ENAMEL POWDER COATING SYSTEMS FOR OVEN CAVITIES WORLDWIDE

Jeff Hale, Phil Flasher
ITW Gema

This presentation will discuss the experiences with an actual production P/E powder spray system.

Successful case studies for powder installations to coat oven cavities with powder porcelain at the following appliance manufacturers will be Electrolux, Neff, Samsung, LG Electronics, Fisher – Paykel, Arcelik, and Bosch Siemens.

Figure 1 shows a photograph of spray guns installed at Neff-Germany to handle the following:

Product:	- Baking ovens
	- Different types of ovens with different sizes in mixed production
Conveyor Speed:	9-18 ft/min
Hanger Distance:	30-36" centers
Coating Process:	in 3 different groups located in 3 booths
Programming:	Touch panel interface with PC control

Figure 1. Powder gun installation at Neff-Germany

Figure 2 shows a similar installations for coating oven cavities at (a) Bosch Siemens in Traunreut, Germany, (b) Electrolux-UK, (c) Fisher-Paykel – New Zealand, and (d) Samsung-Thailand.

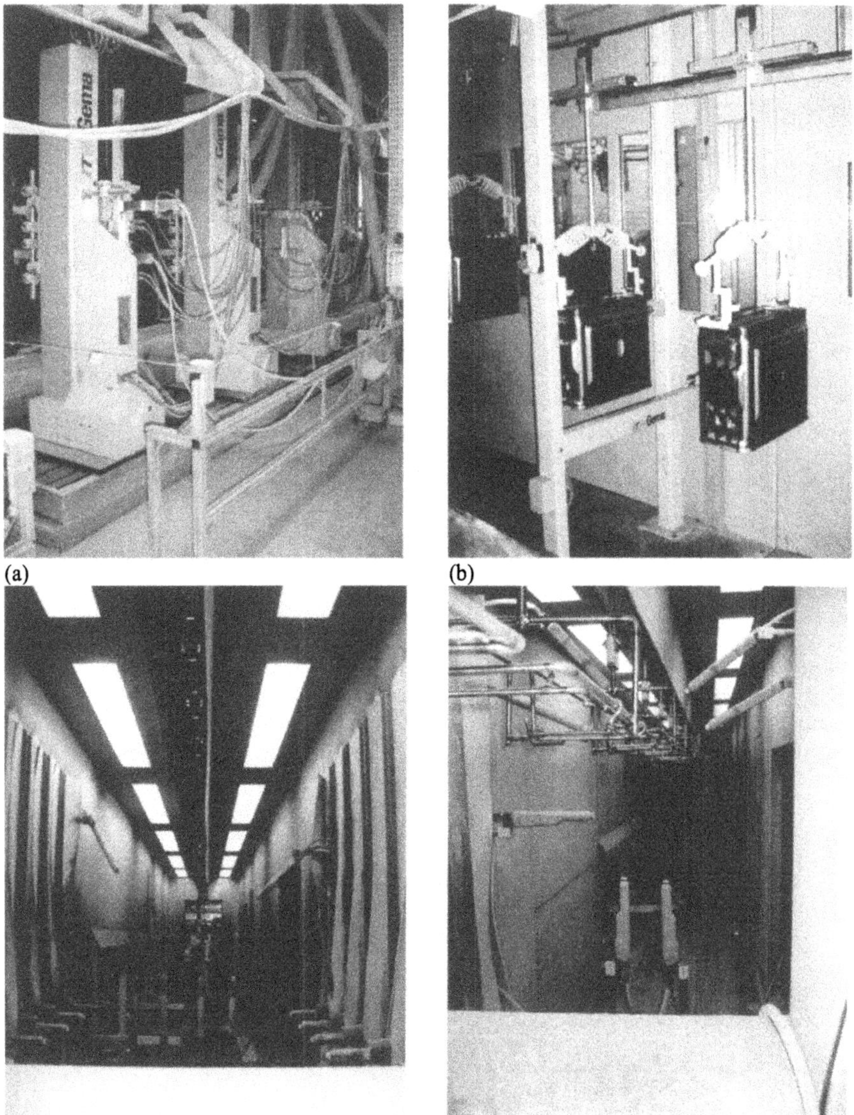

(c) (d)

Figure 2. (a) Bosch Siemens – Traunreut, Germany, (b) Electrolux-UK, (c) Fisher-Paykel – New Zealand, (d) Samsung-Thailand

Figure 3 shows a photograph of newly installed spray guns at LG-Electronics-Monterrey, Mexico.

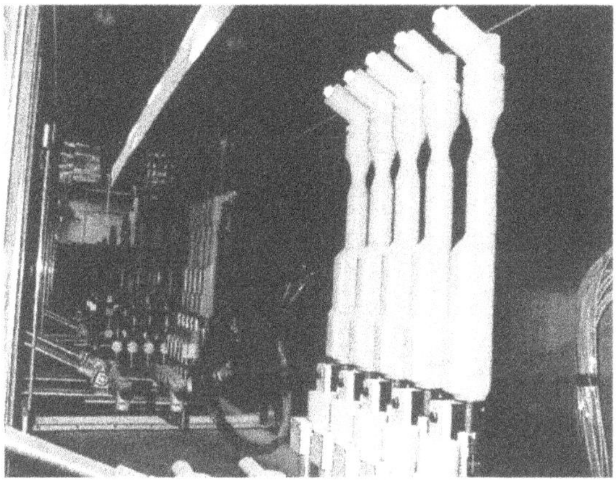

Figure 3. Powder gun installation at LG Electronics-Monterrey, Mexico

As shown by the photos, ITW Gema Technology for porcelain enamel includes application equipment such as guns, controls, and gun movers, as well as recovery booth products like collectors and floor sweeps so powder can be reclaimed, sieved, and fed back into the system.

The ITW Gema OptiFlex automatic gun package has a proven cascade design with a patented volumetric pump airflow control technology. It has a programmable powder output of 0 – 500 g/min in steps of 2% with patented Digital Valve Control (DVC) that controls powder flow rather than air pressure. It gives accurate control of conveying air, supplemental air, and electrode rinsing air. There is digital adjustment for voltage, current, rinse air, powder output and total air. This equipment provides optimum transfer efficiency with reduced orange peel effect from back ionization.

The OptiStar control unit is shown in Figure 4. First, it has separate and exact settings for all coating parameters. Second, it has automatic program changes for different products and different powder outputs can be programmed. This results in consistent powder charging and uniform powder output from all guns. As such, there is constant and consistent build-up of the coating film structure. Continuous compensation of compressed air and dynamic fluctuations are accounted for. This provides a reduction in the average coating thickness, and, therefore, powder savings.

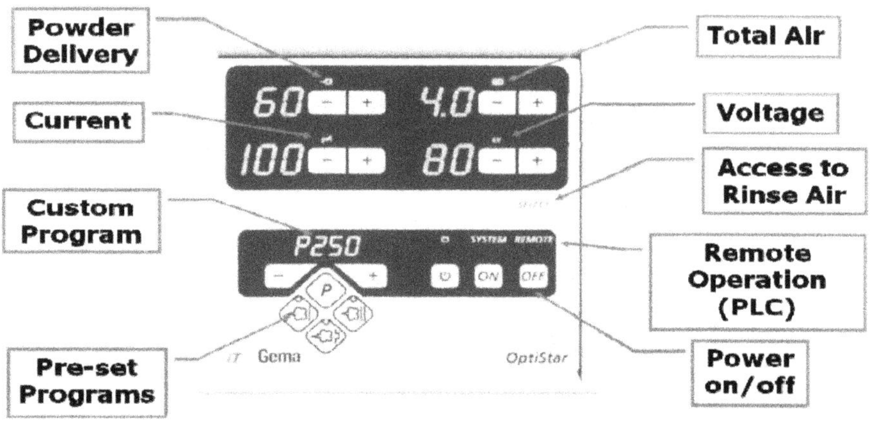

Figure 4. OptiStar control unit

The OptiGun has been specially designed for powder enamel with high charging efficiency by using a flat spray nozzle with a central electrode. A wide range of nozzles for specific parts are available such as a deflector, deflector plate, flat spray, and angle nozzle. The gun consistently and uniformly charges the powder. The current is precisely controlled with voltage adjustment through the OptiStar control unit. A quick-change coupling eases preventative maintenance. The gun has a completely sealed body, which increases safety and gun life. A SuperCorona (ion collection ring) is also available as an option.

The gun movers work in combinations of different axes. This guarantees correct positioning of each individual gun, and the gun movement follows the object size. The gun movers in the installations have the following axes:

- Vertical Z-axis with short or long stroke
- Horizontal X-axis for in and out movement
- Horizontal U-axis for individual gun movement
- Synchronized Y-axis for the conveyor speed related
- Specially designed axis

The gun movers are available as single or multi-axis versions. They are ideal for long or short strokes. All axes are freely programmable with the Optimove and PLC/PC. There is extremely quiet running with the toothed drive belt. These have heavy loading and high stability. Narrow columns permit space saving configuration of the guns. There is a maintenance-free AC motor drive as well.

The automation controls are user friendly and manage the powder output per gun, current limitation per gun, and reciprocator/axis speed and stroke. Programs can be stored for better process control, and the system is connected to Production Process Control system. The graphic interface is shown in Figure 5.

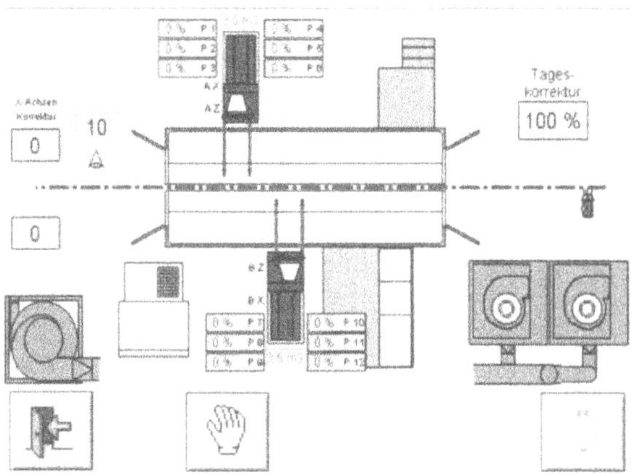

Figure 5. Graphic interface

Figure 6 shows a photo of the MRS-F booth used for enameling oven cavities. To fully utilize recirculated powder, filters specially designed for abrasive porcelain enamel powder are used. The reclaim equipment includes a sieve, heated blanket, and vibrating table.

Figure 6. MRS-F Booth

These ITW Gema installations for powder coating oven cavities result in higher productivity at a lower cost. Repeatable, consistent performance is obtained from reduced material consumption (thinner coat), uniform coating thickness with no drain lines, and the elimination of wet slip rheology issues. Labor is reduced through automation so there is no hand spray and no more wet slip adjustments or beads to remove. Floor space is also reduced because neither a mixing room nor a dryer oven is required. Ovens are coated in a shorter overall cycle time.

DENSE PHASE PUMPING OF PORCELAIN ENAMEL POWDER PART II

Ken Kreeger
Nordson

This paper builds on the 2008 presentation and subsequent experiences with the dense phase pumping of P/E powder.

Nordson HDLV™ Prodigy™ technology moves more powder with less air and higher transfer efficiency than conventional Venturi-style pumps. Figure 1 compares these two types of spray systems.

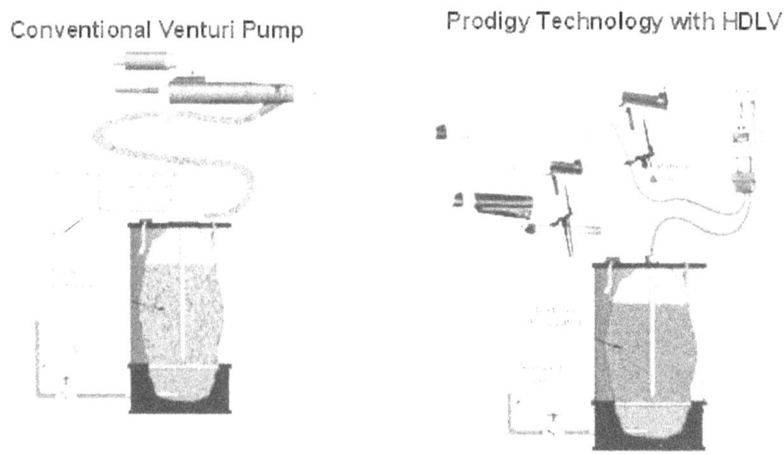

Figure 1. Comparison of standard and HDLV powder guns

The core of this new powder application technology is the dense phase powder pump, which utilizes high-density powder with low-velocity air as shown in Figure 2. The HDLV™ pump resembles a human heart in that it consists of two side by side pumps each with an upper and lower chamber working together to provide precise, carefully synchronized powder transport.

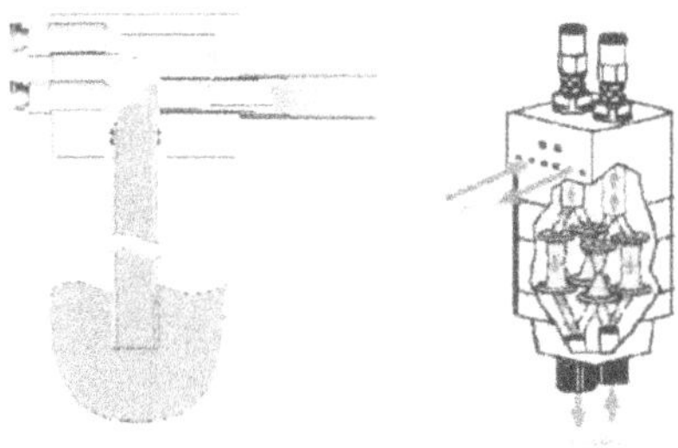

(a) (b)

Figure 2. (a) Venturi pump and (b) HDLV™ pump

A more detailed schematic of the HDLV™ pump is shown in Figure 3. The spray pattern is formed independently of powder transport. Absolutely minimal air volume is used for powder delivery.

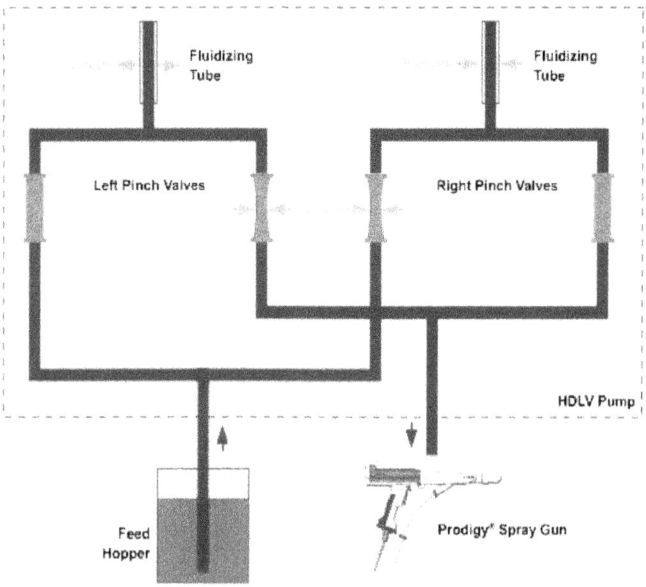

Figure 3. HDLV™ pump schematic

The Prodigy™ system optimizes transfer efficiency by balancing the air volume control with the HDLV™ pump. Historically most of the focus has been on electrostatics and charging system controls. Factors affecting corona charging are shown in Table 1.

Factor	Comments
Charging Time (τ)	Gun-to-part distance Powder velocity Electrode/Nozzle configuration
Field Strength (E)	Electrostatic parameters (kV: mA) Gun-to-part distance Electrode/Nozzle configuration
Particle Size (a)	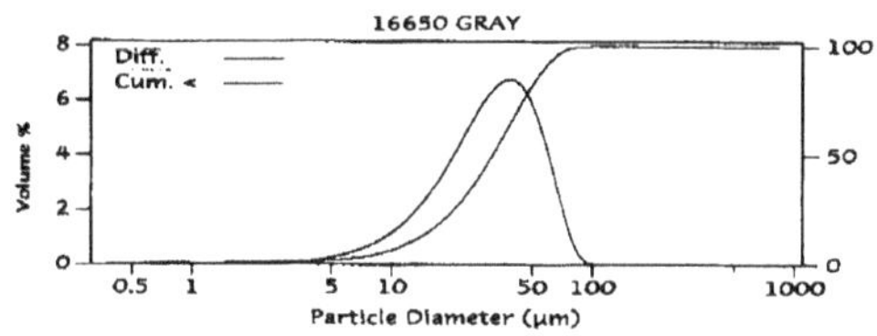

Table 1. Factors affecting Corona charging

Figure 4 shows the impact of different variables on the porcelain enamel powder process.

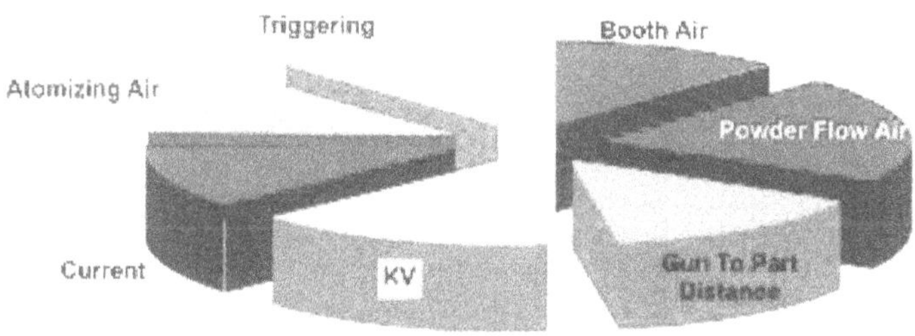

Figure 4. Porcelain powder process variables

To optimize the aerodynamics of a powder application process, several issues must be addressed: (1) the spray pattern dynamics of an applicator, (2) the working characteristics of a powder pump, (3) the powder-to-air ratio, (4) aerodynamic turbulence in the powder deposition area.

Aerodynamics play important role in powder application efficiency and coverage of recessed areas. The application of powder with a Venturi pump-type gun is shown in Figure 5. The powder pump is designed to deliver powder to the applicator in a controlled, consistent, and uniform flow. However, the relatively high velocity of the powder can cause it to rebound from the ware, making it difficult to coat recesses. Furthermore, all powder-contact surfaces in the Venturi pump are made of wear-resistant ceramic materials because of the abrasive nature of high velocity glass powder particles.

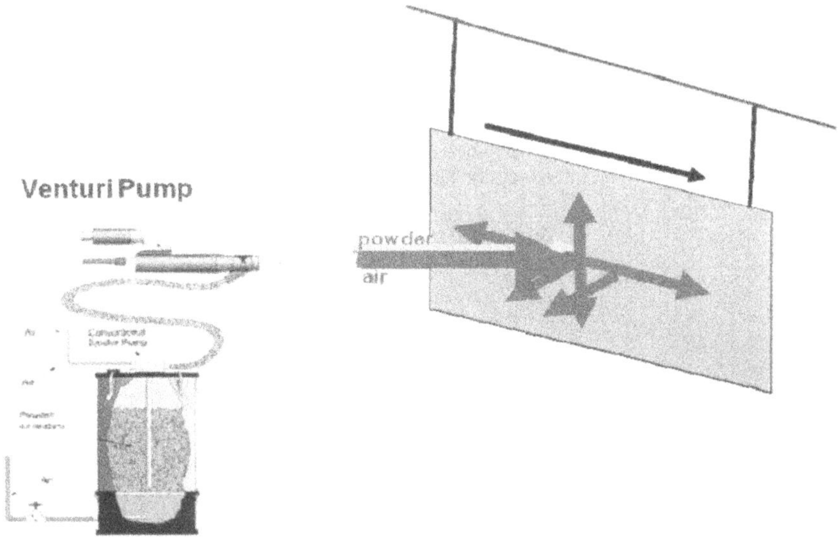

Figure 5. Powder aerodynamics with a Venturi pump

Figure 6 compares the hoses for a standard powder gun to an HDLV™ gun. With a Venturi pump, the hose is 60% air. With an HDLV™ pump, the hose is more packed with powder and a smaller inner diameter (ID) is used.

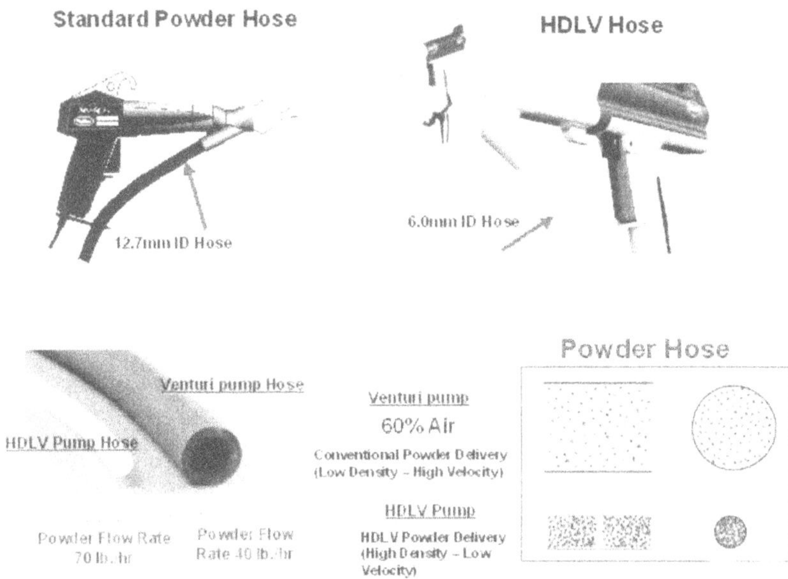

Figure 6. Comparison of powder pump hoses

With the HDLV™ gun, the powder cloud has a high concentration of powder with very little air and a softer spray pattern as shown in Figure 7. With the softer spray pattern, the gun-to-part distance can be reduced from 8 to 10 inches down to 3 to 6 inches. The powder velocity is also 5 times slower.

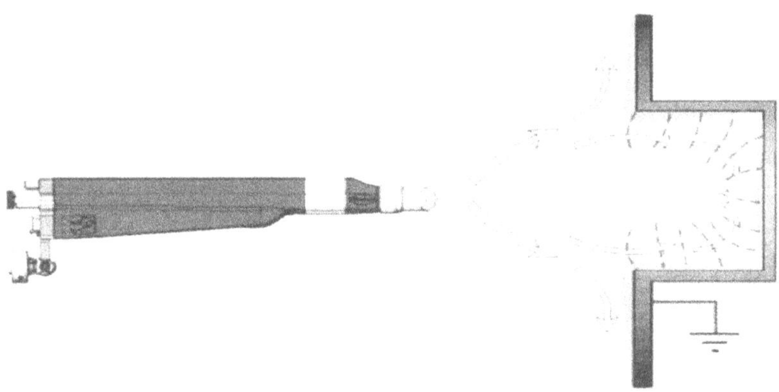

Figure 7. Softer powder cloud with HDLV™

The Prodigy HDLV™ Pump has no moving parts and is designed to purge in both directions and color change quickly. The design is see through for easy diagnostics. The pump has an output rate of 27 kg/hr through a delivery hose with an ID of 6 mm.

The manual Prodigy system uses a controller specialized for recoating, touch-up, and complicated parts with ten presets for user convenience. This is designed for fast color change and high first pass transfer efficiency.

The Prodigy™ HDLV™ automatic gun controller uses closed loop digital gun control, part detection and gun triggering, gun mover control, and can store 255 coating recipes at one time. The application controller runs up to 4 reciprocators with different mode options, has control of up to 4 in / out movers with different mode options, stores reciprocator parameters in 255 coating programs, and flags recognition for automatic preset loading. A schematic of the automatic Prodigy™ gun is shown in Figure 8. It is used with iControls software. The Prodigy II HDLV high-capacity pump can be used for virgin or reclaim powder feed. It utilizes less compressed air, delivers three times the powder of a venturi-style transfer pump, and speeds change over time.

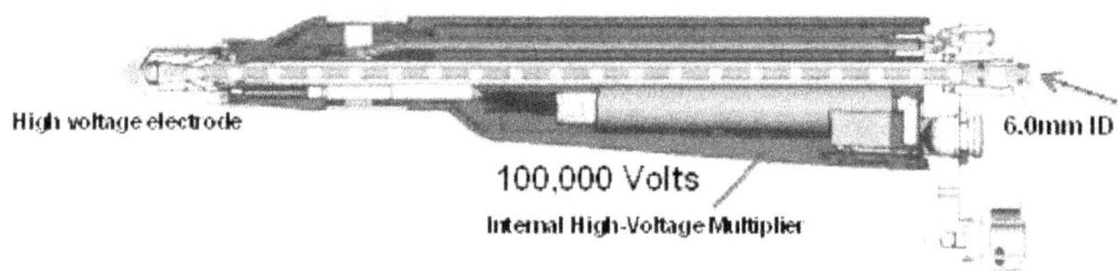

Figure 8. Prodigy™ gun schematic

Figure 9 shows a photograph of a robot mounted Prodigy™ gun used to evenly coat the interior of an oven cavity. With a Venturi pump, the powder moves too fast and does not penetrate recessed Faraday cages. The softer powder cloud produced by an HDLV™ pump has much better coverage of recessed areas so enameling oven cavities is a perfect application of the Prodigy™ gun.

Figure 9. Photograph of Prodigy HDLV Robot PE System

The HDLV™ pump can also improve powder transport. A Venturi Pump has a capacity of 120 lbs/hr non-fluidized or 130 lbs/hr fluidized using 7 to 10 scfm. The pump capacity of the HDLV™ pump is 300 to 360 lbs/hr non-fluidized or 500 to 600 lbs/hr fluidized using only 2 to 7 scfm. As such, the HDLV™ technology delivers better powder transfer as shown in Figure 10. One to two HDLV™ pumps can move as much powder as 3 to 5 Venturi pumps. It also has simple pneumatic control for ease of operation. Less compressed air used to propel the powder results in more efficient recycling. Large amounts of powder are transported in a short amount of time. Finally, the system is automatically purge-able with either continuous or pulse purge.

Figure 10. Drum unloader – virgin powder delivery

The Prodigy HDLV gun pump moves powder to the spray gun with low velocity air. This results in softer delivery, minimal overspray, higher transfer efficiency, reduced usage of compressed air, and greater material savings. There is less powder to move throughout the system, which improves productivity and the consistency of the coating. The pumps and spray guns are automatically purgeable in both the suction and delivery directions for faster clean up.

The HDLV™ powder transfer system uses less compressed air to propel the powder through the recovery system. This has smaller powder transport tubing and also less compressed air usage. The consistency of the powder porcelain is improved, and downtime is reduced. This reduces labor costs and gives faster change over times. Most importantly to enamellers, the Prodigy™ can coat recessed areas very uniformly.

CASE STUDY: PROCESS RHEOLOGY OF COVER COATS AT ENGINEERED STORAGE PRODUCTS COMPANY

Greg Budnick
Engineered Storage Products Company (ESPC)

This paper will review the experiences at ESPC with control of porcelain enamel slip rheology for wet spray systems.

Engineered Storage Products Company (ESPC), located in DeKalb, IL, manufactures porcelain enameled steel, bolt-together tanks. The tanks have agricultural, municipal, and industrial applications.

Porcelain enamel is milled on-site. The storage tanks are both water-jacketed and non-jacketed. One need was to improve the cover coat by optimizing milling time. The milled enamel needs to come off with a pick-up of 50 g/ft^2 which is then adjusted to 100 g/ft^2. The initial set is lowered to allow material to go through the Roto Spray.

For the first project, the goal was to improve the chemical resistance of the cover coat. A cover coat with enhanced corrosion resistance was substituted into the original formula. The original and improved formulas are shown in Table 1.

Raw Material	Original	Improved
Frit	100%	100%
Clay	5%	5%
Bentonite	0.25%	0.25%
Sodium Nitrite	0.062%	
Magnesium Carbonate	0.062%	
Silicon Dioxide	4%	5%
Titanium Dioxide	4%	0.25%
Tetrasodium Pyrophosphate ("Pyro")	100g	
Potassium Carbonate		0.062%
Monopotassium Phosphate (MKP)		0.062%

Table 1. Cover coat formulas

The improved formula was first tested in the company lab in preparation for production line trials. The first production mills came off with low set. Incremental changes were made so as not to drastically change the formula. These are summarized in Table 2. Strong support was received from the frit supplier during the production line trials.

Change	Result
Added more clay	High initial set
Changed clay	High initial set, unable to adjust further to obtain desired set point
Sodium Aluminate	Too much initial set, difficulty implementing as a set up agent
Pyro w/ all of above	Reduced set, but continued loss of set upon ageing
Pot. Phosphate	Lower set out of mill, but more stability upon ageing
Pot. Phos. w/ Pot. Carb.	Raised the set point and maintained stability

Table 2. Summary of mill addition changes

The next issue encountered with the formula was the pick-up, which was 75 g/ft^2 compared to 45 to 60 g/ft^2, which was required for the Roto-Spray. Forty grams of pyro was added post milling to correct this.

For the second project, the blue ground coat formulation was successfully used to make a green cover coat. This was achieved by adding additional oxides to original blue glass formula. While the blue glass worked well, green had ageing problems.

Possible causes of green slip ageing problems are no cooling jackets on green storage tanks, the green not consumed as fast as the blue, or the green enamel being more sensitive to the set-up salts. As of this writing, testing on the green glass is in the beginning stages. Favorable results have been obtained by adding clay and bentonite to the formula, but more work needs to be done to achieve significant cost reductions by eliminating green glass resprays and moving to single-fire processing.

Acknowledgements

Many thanks to Jason Butz, Miles Votava, and Larry Wulf for their time and support in compiling this paper, since I was not working at ESPC when most of this work was completed.

WATER HEATER AND INSIDE COATING WITH WET ENAMEL, USABLE FOR SOLAR ENERGY

Brigitte Riester-Alt
E.I.C Group GmbH

New application system technologies for porcelain enamel for solar energy collectors are discussed.

Figure 1 shows a water heater that uses a solar collector. There are several reasons why this unit is made with enameled steel rather than stainless. First stainless steel is costly--a mild steel tank with enamel coating is only 60% of the cost of a stainless steel tank. Second, water quality is getting worse and deteriorates stainless steel, which results in more warranty claims from customers. Third, there is a desire to offer a longer warranty period, and with enamel coated mild steel it is possible to have a guarantee of more than 10 years with lower cost of claims during warranty period. Fourth, the overall production rates are increasing for water heaters with internal pipes which can be met with wet application techniques with fully automated lines. Starting in the late 90s, investments have been made in water heater manufacturing that were forward looking and planned to encompass current economical and ecological aspects, especially for the use of solar energy.

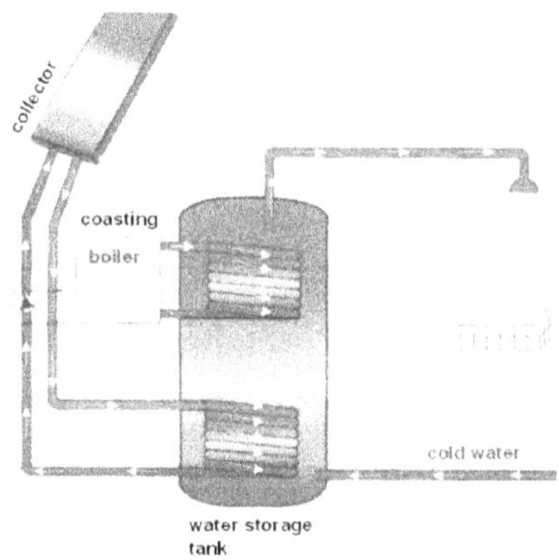

Figure 1. Solar water heater

Approximately 70% of European water heaters now have internal pipes. The water heater size is between 80 and 500 L, with an average between 200 - 300 L. The minimum production capacity rose above 800 water heaters/7.5 hour shift, with some installations having an average of more than 2.500 water heaters/7.5 hour shift. The automation of the coating process corresponded to the wide range of water heater shapes, as well as the diameter and the length. Automation to provide maximum flexibility of the system was generally purchased and installed.

Figure 2 shows the method of hanging the water heaters during production that has met the production requirements.

Figure 2. Economical water heater loading concept

Figure 3 shows an example of the internal pipe shapes that need to be enameled in mass production. Most of the tasks are fully automated, with three exceptions. First, workers still clean the pipe outlet connections. Second, operators still set specific parameters for inside coating of the water heater. Third, workers are still loading and unloading the water heaters.

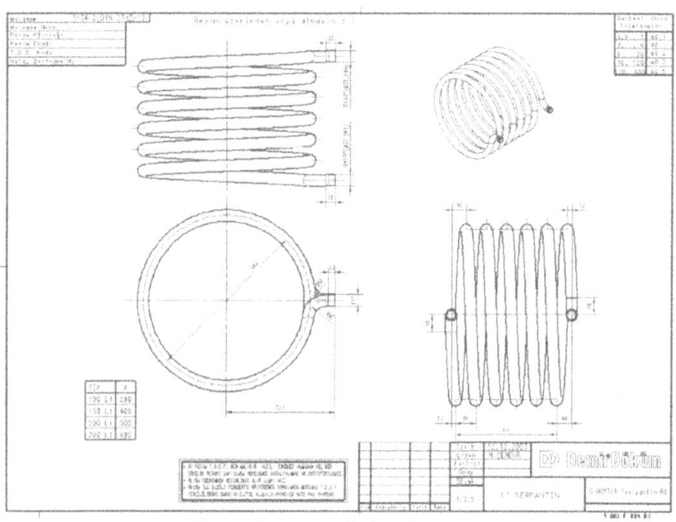

Figure 3. Internal pipe shapes

Figure 4 shows the enamel system layout used. There is 100 % recovery of excess enamel, which reduces waste water, and there is no airborne dust.

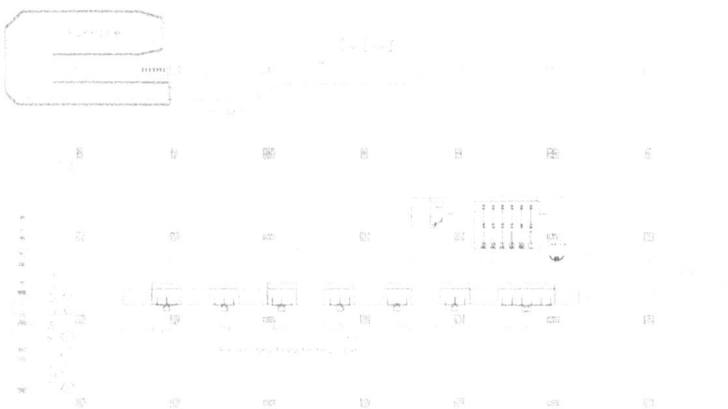

Figure 4. Enamel system layout

Figure 5 shows the ready to use enamel delivered by various enamel manufacturers in big bags. This avoids the use of mills and considerably improves the environmental aspect of the wet application by using less fresh water.

Figure 5. Big bag system

The interior of the water heater is flow coated, and the interior pipes are coated with a high fluid volume transfer system. A photograph is shown in Figure 6.

Figure 6. Fully automated water heater inside coating line with and without internal pipes

The production time for one water heater with its transport, swiveling, flooding, draining and distribution is dependent on the process time required. This is determined by the water heater's size, diameter of heating pipes, position, construction and distances of pipes, the slip material adjustment, and the required number of coatings for thickness and quality.

In summary, the quality of the water heaters has been improved at reduced production costs. There is less material waste using reliable fully automated production equipment with low cost maintenance and the latest technology. All provide investment savings while contributing to the efficient use of the solar water heaters.

SUSTAINABILITY – MORE THAN A BUZZ WORD

W. Mark McGinley, Ph. D., PE
University of Louisville

A.I. Andrews Memorial Lecture

This paper explores the sustainability issue as it relates to construction and materials in the 21st century with particular emphasis on the impact to manufacturers.

Sustainability will be defined as it relates to building design. The selection of practical sustainable green materials using Life Cycle Assessment (LCA) will be reviewed as well as opportunities and challenges to meet green/sustainable design criteria.

Why sustainability? The first reason, shown in Figure 1, is that, with birth rates sharply exceeding death rates, particularly in the developing world, population growth rates have been very high and there are more people than ever.

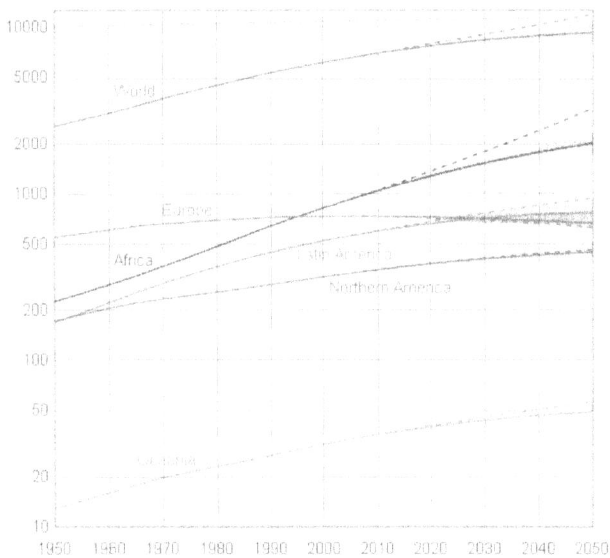

Figure 1. Population growth rates (UN)[1]

The second reason is that there is increasing competition for fewer resources. Figure 2 shows Hubbert's curve for oil. Peak in the curve for the US occurred in ~1970, and the peak in the curve for global production is around now while the peak in the curve for the Middle East has not yet occurred.

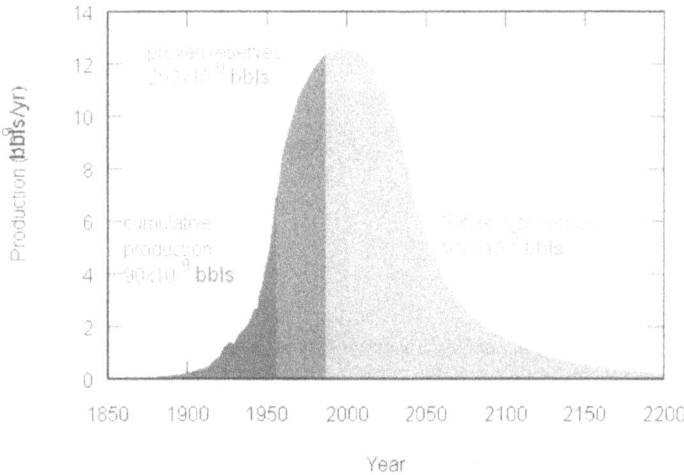

Figure 2. Hubbert's curve for peak oil[2]

Regardless of global warming and climate change, it is clear more people will compete for fewer resources, which will ultimately run out. Until alternatives are developed, existing resources need to be stretched as much as possible. It is the right thing to do and ultimately will be a good business decision.

As defined by the World Council on Environment and Development (WCED) in 1987, sustainable development is that which "meets the needs of the present without compromising the ability of future generations to meet their own needs."[3] Sustainable designs include social, economic, and environmental criteria. Generally designs of systems must meet needs and do so in the most cost effective, safest, aesthetically pleasing and durable manner possible. As such, incorporating sustainability in design adds more criteria to the mix. Until now efficiency was the prime criteria, and everything was framed in terms of cost. Since environment and social criteria have been added, cost is not always best way to compare alternatives.

For construction, sustainable design for buildings facilities in the United States has two counting methods:

- NEED: LEED (leadership in Energy and Environmental Design)
- US Green Building Council –www.usgb.org

LEED has many guidelines for the main areas of sustainable facility design such as a rating for a new designs and operations that evaluates the environmental performance of systems and the whole building over its lifecycle. LEED requirements are broken down into six categories and are assigned a point rating system as follows:

1. Sustainable Sites (14 pts.)
2. Water Efficiency (5 pts.)
3. Energy & Atmosphere (17 pts.)
4. Materials & Resources (13 pts.)
5. Indoor Environmental Quality (15 pts.)
6. Innovation & Design Process (5 pts.)

For new construction, LEED is rating as follows:

- Certified 26–32 points
- Silver 33–38 points
- Gold 39–51 points
- Platinum 52–69 points

Figure 3 shows examples of LEED sustainable sites.

Yes	?	No			
			Prereq 1	**Construction Activity Pollution Prevention**	Required
			Credit 1	**Site Selection**	1
			Credit 2	**Development Density & Community Connectivity**	1
			Credit 3	**Brownfield Redevelopment**	1
			Credit 4.1	**Alternative Transportation**, Public Transportation	1
			Credit 4.2	**Alternative Transportation**, Bicycle Storage & Changing Rooms	1
			Credit 4.3	**Alternative Transportation**, Low-Emitting & Fuel Efficient Vehicles	1
			Credit 4.4	**Alternative Transportation**, Parking Capacity	1
			Credit 5.1	**Site Development**, Protect or Restore Habitat	1
			Credit 5.2	**Site Development**, Maximize Open Space	1
			Credit 6.1	**Stormwater Design**, Quantity Control	1
			Credit 6.2	**Stormwater Design**, Quality Control	1
			Credit 7.1	**Heat Island Effect**, Non-Roof	1
			Credit 7.2	**Heat Island Effect**, Roof	1
			Credit 8	**Light Pollution Reduction**	1

Figure 3. LEED sustainable sites

Figure 4 shows examples of LEED materials and resources. Note the emphasis on post-consumer recycled content.

Yes	?	No		Materials & Resources	13 Points
Yes			Prereq 1	**Storage & Collection of Recyclables**	Required
			Credit 1.1	**Building Reuse,** Maintain 75% of Existing Walls, Floors & Roof	1
			Credit 1.2	**Building Reuse,** Maintain 95% of Existing Walls, Floors & Roof	1
			Credit 1.3	**Building Reuse,** Maintain 50% of Interior Non-Structural Elements	1
			Credit 2.1	**Construction Waste Management**, Divert 50% from Disposal	1
			Credit 2.2	**Construction Waste Management**, Divert 75% from Disposal	1
			Credit 3.1	**Materials Reuse**, 5%	1
			Credit 3.2	**Materials Reuse**, 10%	1
			Credit 4.1	**Recycled Content**, 10% (post-consumer + 1/2 pre-consumer)	1
			Credit 4.2	**Recycled Content**, 20% (post-consumer + 1/2 pre-consumer)	1
			Credit 5.1	**Regional Materials**, 10% Extracted, Processed & Manufactured	1
			Credit 5.2	**Regional Materials**, 20% Extracted, Processed & Manufactured	1
			Credit 6	**Rapidly Renewable Materials**	1
			Credit 7	**Certified Wood**	1

Figure 4. LEED materials and resources

LEED is changing and being streamlined into three general categories totaling 100 points with more emphasis on energy use and regional issues.

The Green Globes Green Building Initiative (GBI) is based on the United Kingdom (UK) Building Research Establishment's Environmental Assessment Method (BREEAM), which was started in Canada in cooperation with ECD Energy and Environment. In 1996, the Canadian Standards Association published BREEAM Canada for Existing Buildings. Green Globes provides a rating of environmental impact of new commercial buildings and major renovations using an assessment protocol that covers the seven different areas shown in Table 1. The scoring for these seven categories is based on questions completed via the online questionnaire within the Green Globes Tool.

Assessment Category	Points	Description
Energy	360	Performance, efficiency, demand reduction, energy efficient features, use of renewable energy, transportation
Indoor Environment	200	Ventilation, lighting, thermal & acoustical comfort, ventilation system
Site	115	Ecological impact, development area, watershed features, enhancement
Resources	100	Low impact materials (LCA), re-use, demolition, durability, recycling
Water	100	Performance, conservation, treatment
Emissions & Effluents	75	Air emissions (boilers), ozone depletion, water & sewer protection, pollution controls
Project Management	50	Design process, environmental purchasing, commissioning
Total	1000	

Table 1. Green Globes rating system

For the selection of systems and materials, environmental and societal criteria need to be met. Systems and materials need to be selected that provide minimum impact on the environment globally, regionally and locally. Societal effects can be addressed by creating a healthy interior/exterior environment, minimizing pollution and use of nonrenewable resources.

There is a need to have certified "Green Systems", and there are organizations doing this such as the Blue Angel Label in the Nordic countries and the Swan in Germany. The US forestry industry has the Forest Stewardship Council (FSC) certified managed sustainable forest products. There are EPA and Euro standards that look at green products as those that have a minimal impact on indoor or outdoor environments. The US government has what is referred to as Environmentally Preferred Products and has a scientific method to address environmental effects called a Life Cycle Assessment (LCA).

What is meant by "Green" materials varies with use. As such, it can be unclear what a green material or system is. Many products or systems can be called "green" depending on what you mean by "green" and its use so it can be difficult to compare systems and materials with these variations in definitions and certifications.

One solution is the Life Cycle Assessment (LCA). This is a scientific method for assessing the resources, emissions, other environmental impacts of a material or system from extraction through disposal. This process consists of several steps defined in the ISO 14000 series of standards. LCA addresses inventory analysis, impacts assessment, and interpretation of the impacts.

Specifically LCA accounts for:

- Fossil fuel depletion
- Other nonrenewable resource use
- Water use
- Global warming potential
- Stratospheric ozone depletion
- Ground-level ozone (smog) creation
- Nutrification eutrophication of water bodies
- Acidification and acid deposition (dry and wet)
- Toxic releases to air, water, and land

One LCA program in use is ATHENA, which is used in Canada and the United States, while Building for Environment and Sustainable (BEES) is only in the US.

ATHENA focuses on whole buildings or building assemblies such as walls, roofs, or floors. It addresses material maintenance and replacement over an assumed building life. Energy generation simulation results are entered into a database that covers 90 structural and envelope materials and can simulate more than 1,000 different assembly combinations. An example from the literature is shown in Table 2.

Example of an LCA Output

Building Components	Embodied Energy (GJ)	Solid Waste (metric tons)	Air Pollution* (index)	Water Pollution* (index)	GWP† (equivalent CO_2 metric tons)	Weighted Resource Use (metric tons)
Structure	52,432	3,273	859.0	147.0	13,701	34,098
Cladding	17,187	281	649.8	24.7	5,727	2,195
Roofing	3,435	145	64.8	5.8	701	1,408
Total	73,054	3,554	1,573.6	177.5	20,129	37,701
Per square meter	2.36	0.11	.05	.006	.65	1.21

*The air and water pollution indices are based on the critical volume measure (method).
†GWP is global warming potential. Energy and emission estimates do not include operating energy.

Table 2. Example of an LCA output

Figure 5 shows the output screen from ATHENA with the rating system.

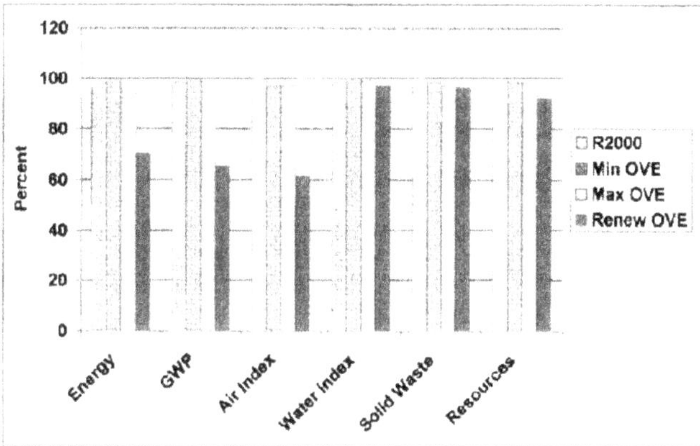

Figure 5. ATHENA output

BEES measures the environmental performance of building products by using the life-cycle assessment approach specified in the ISO 14040 series of standards.[4] All stages in the life of a product are analyzed: raw material acquisition, manufacture, transportation, installation, use, and recycling and waste management. Economic performance is measured using the ASTM standard life-cycle cost method, which covers the costs of initial investment, replacement, operation, maintenance and repair, and disposal. Figure 6 shows example of BEES output comparing different types of cement for combined economics and environmental performance. The three are similar on economics, but quite different on environmental impact.

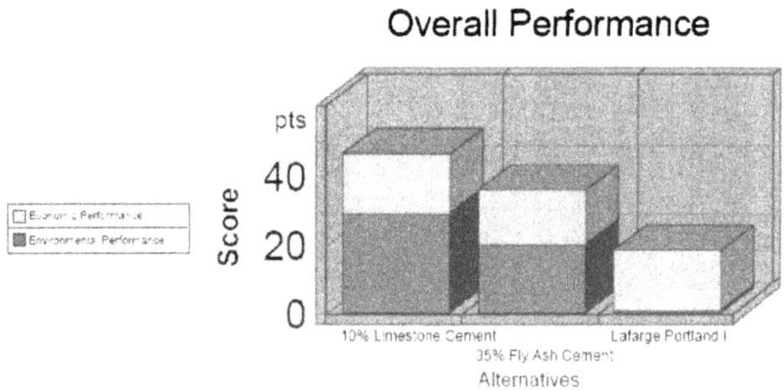

Note: Lower values are better

Category	10% Lime	35%FlyAsh	Lafarge I
Economic Perform.--50%	17.2	15.5	17.2
Environ. Perform.--50 %	29.0	20.1	0.9
Sum	46.2	35.6	18.1

Figure 6. Exampe of a BEES result

There are still problems with the LCA analysis method. The environmental impacts are integrated using methods some find inflexible, and it is hard to compare across categories. While LCA programs are easy to use, the materials database is limited. LCA may not show the whole story driving the material selection. For example, a critical design criterion like fire resistance may be required. LCA may not be prefect but is still being promoted and used (along with LEED and Green Globes).

Energy intensive processes do not come out well in LCA analysis. Steel has poor LCA analysis, even though the majority of it is recycled. Concrete faces similar issues. If one uses a lot of process energy, there is not a reason for panics there are other ways to compare sustainability of systems or materials.

Charles Kibert has suggested three other ways of looking at what is a truly environmentally friendly "sustainable" material[5]. First, the Natural Step framework developed by Dr. Henrik Robert is used which sets out four conditions that overall required materials used are non-persistent, nontoxic and procured either from reused, recycled, renewable, or naturally abundant sources.[6] Second, the Cardinal Rules for Closed-Loop Building Materials needs to be followed so the materials and systems must form a closed loop like nature with near zero waste. Buildings must be deconstructable, products must be disassemblable, materials must be recyclable, products materials must be harmless in production and in use, and materials dissipated from recycling must be harmless. Third, the "Pragmatic View of Green Building Materials", described in the Environmental Building News (EBN), breaks green building products into five major categories with a number of subcategories.

The first category are environmentally attractive material like salvaged products, materials, those with postconsumer and/or postindustrial recycled content, certified wood products, rapidly renewable products (<10 years), those made from agricultural waste material, and minimally processed products. Second, products that are green because of what is not there like ones that reduce material use as well as those that do not contain ozone-depleting substances, PVC, polycarbonate, conventional preservative-treated wood, or hazardous components. The third category are products that reduce environmental impacts during construction, renovation, or demolition as well as those that reduce the impacts of new construction, renovation, or demolition. Fourth, products reducing the environmental impacts of building operation like heating and cooling loads or equipment that conserves energy and so forth. Finally, the fifth category is products that contribute to a safe and healthy indoor environment by not releasing significant pollutants into the building, blocking the introduction, development. or spread of indoor contaminants, removing indoor pollutants, warning occupants of health hazards in the building, or improving light quality.

So what a "green" or "sustainable" material is unfortunately depends on the definition or measurement used. Unfortunately, energy intensive products like porcelain enamel are under pressure no matter what scale is used. However, challenges can be transformed into opportunities. Energy use in the process can be reduced or renewable energy can be incorporated. Water use, waste stream, and pollutants can be reduced, and as much recycled material as possible, preferably post-consumer glass, needs to be used. Perhaps frit systems and processes can be developed that are readily recycled, or the enamel could be removed with less energy. Recycling industries could also be developed to process enamel materials and post-consumer glass. Finally, LCA and environmental analysis is relative. One only needs to be better than the competition. For example, porcelain enamel could replace less environmentally-friendly coatings like galvanization.

Sustainability is here and will become increasing important in design and material selection, particularly as it is legislated. Doing more with less with minimal impact on the environment will be mandated. Opportunities for diversification and market are out there with new processes, systems and products. Eco-performance is relative.

References

[1] "Overpopulation"
http://en.wikipedia.org/wiki/Overpopulation
(2 December 2009)

[2] "M. King Hubbert"
http://en.wikipedia.org/wiki/M._King_Hubbert
(2 December 2009).

[3] "Report of the World Commission on Environment and Development: Our Common Future – A/42/427 Annex"
http://www.un-documents.net/wced-ocf.htm
(2 December 2009).

[4] "BFRL: Office of Applied Economics"
http://www.bfrl.nist.gov/oae/software/bees

(2 December 2009).

[5] Herman E. Daly with Charles J. Kibert, Ed., *Reshaping the Built Environment* (Island Press, 1999).

[6] "The Natural Step"
http://en.wikipedia.org/wiki/The_Natural_Step
(2 December 2009).

DECORATING WITH CERAMIC DECAL

Michael J. McCall
Heinrich Ceramic Decal, Inc.

ABSTRACT

Decorating porcelain enamel substrates can prove exceptionally challenging when trying to reproduce fine detailed images on irregular shapes. Ceramic decal can be used as an alternative decoration method when other decorating techniques are not practical.

HOW CERAMIC DECALS ARE MADE AND TRANSFERRED

A ceramic decal is a transfer that is comprised of metallic oxides that are screen printed onto a specialized paper. Starting with digital artwork, the art is separated into the various colors required to reproduce the image. An application such as Adobe Photoshop is commonly used. The software separates the artwork into individual colors that will be recombined during the printing process (Fig. 1).

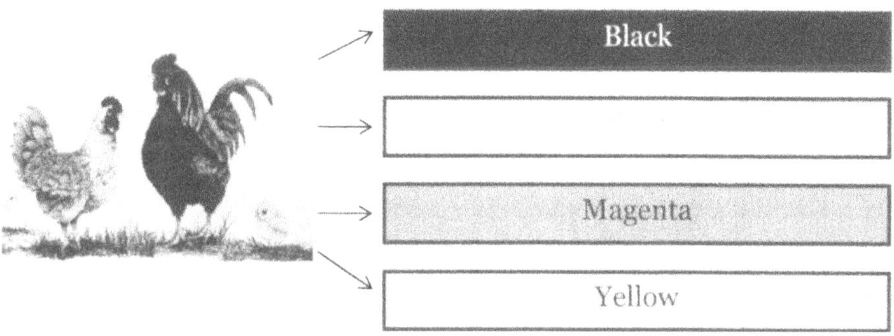

Figure 1. Color Separation

Once the color separation is complete, the separation data is sent to either an imagesetter, to produce film positives which are imaged onto a screen, or to a computer-to-screen imaging device that directly images the screens. Each color will be imaged on its own screen.

Ceramic decal is comprised of three main components: paper, inks and covercoat. Decal paper is used as the base carrier to print all the required colors. The inks are similar to those used in direct screening for porcelain enamel. Metallic oxides are mixed with various fluxes and medium to make a printing paste. Decal inks are generally formulated to dry faster, since heat is not used in the drying process. The last component of a decal is the covercoat. Covercoat is an organic material made from solvent and acrylic resins. The function of the covercoat is to hold the ceramic inks together as the decal is transferred from the paper to the porcelain substrate. Since the covercoat is organic, it burns off during the firing process along with the medium from the ink.

There are two types of ceramic decal paper: waterslide and heat release. Waterslide paper starts with a base paper that is designed to readily absorb water. Eucalyptus pulp has been commonly used for base paper. The base paper is coated with a layer of clay to provide a smooth print surface. Next a layer of starch is applied. The starch becomes very slippery when wet, allowing the decal to slide off of the paper. The final layer of coating is dextrin gum. The gum stays on the bottom surface of the decal after transfer and serves as an adhesive to form a temporary bond between the decal and the substrate before firing. Since the dextrin gum is organic, it too burns off during firing.

Heat release paper generally uses the same base paper as waterslide paper, but is coated differently. Instead of starch and gum, a thin layer of wax is coated onto the base. This wax has a low melting point, somewhere just over 100°F (38°C).

There are many variations of ceramic decals. Waterslide decals are used for manual decoration of small quantities. Heat release decals are needed when decorating large volumes of a substrate with a consistent shape.

Ceramic decals can be fired at various temperatures ranging from 1400-2200°F (760-1200°C). Decals fired between 1400-1600°F (760-871°C) on the surface of a glaze are known as onglaze decals. Decals fired between 1600-1800°F (871-982°C) into the surface of a glaze are known as inglaze decals. Inglaze decals are more durable since they are set into the glaze by the higher temperature firing. Decals that are fired between 2000-2200°F (1093-1200°C) are called underglaze decals. These decals are applied before a glaze is applied to a substrate and offer the best durability.

Glass decals use the same basic ceramic colors or stains, but are mixed with fluxes that have lower melting points. Glass decals fire in the temperature range of 1000-1200°F (537-650°C).

Porcelain enamel decals can fall in between the traditional glass and ceramic decals. Colors are custom blended to work in various flash firing cycles that are sensitive to the enamel coating. Each decal must be fine tuned to work properly in different firing processes.

Decals can be printed in flat sheets for waterslide or heat transfer applications. Flat sheet heat transfers are used in machines with specific magazine sizes or in semi-automatic application processes. Decals printed on rolls can be used in direct and indirect heat transfer machines.

A waterslide decal is built by screen printing multiple layers of ceramic colors onto transfer paper and then coating the entire image are with a layer of covercoat (Fig. 2). To transfer the image, the decal is first soaked in water for 30-40 seconds. Water is absorbed through the back of the porous base paper, which then softens the starch layer and wets the dextrin gum. After the decal is removed from the water, the image is slid off of the transfer paper and applied to the substrate. A squeegee is used to remove all air and water from underneath the decal. The substrate is allowed to dry thoroughly, preferably overnight, and then fired.

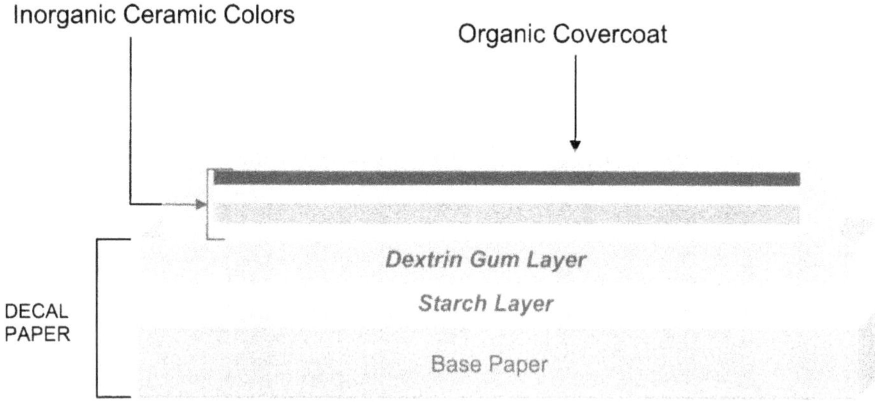

Figure 2. Composition of a Waterslide Decal

A heat release decal is printed in the same manner as a waterslide decal, using the wax coated paper. If the decal will be used as an indirect transfer, the image is printed right reading. A covercoat is used to bind the ink together for the transfer process. If the decal is applied as a direct transfer, the image must be printed wrong reading. A direct transfer has a layer of heat activated adhesive printed instead of the covercoat layer on top.

The indirect transfer process used a heated silicone pad that comes into contact with the top surface of the decal (Fig. 3). The covercoat and ink are transferred to the pad. The pad is then pressed against the substrate to transfer the ink to the substrate. The substrate may be fired immediately after transfer occurs.

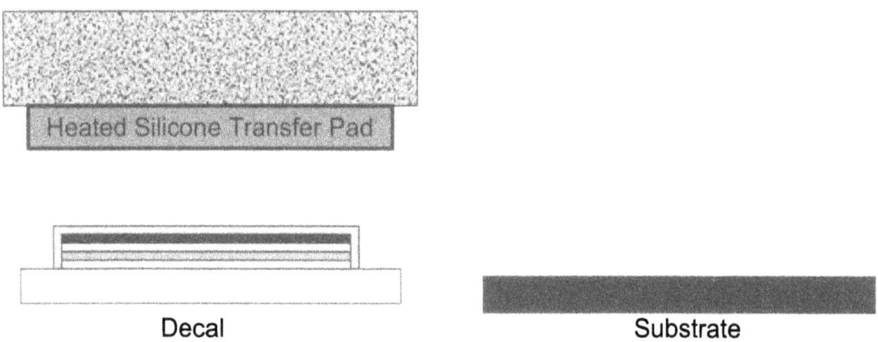

Figure 3. Transferring an Indirect Heat Transfer Ceramic Decal

The direct transfer process applies heat to the bottom side of the decal (Fig. 4). The heat is absorbed through the decal paper, melting the layer of wax. The decal is pressed against the substrate with the heated pad. Since the wax has melted, the ink is easily transferred to the substrate. The substrate may be fired immediately after transfer occurs.

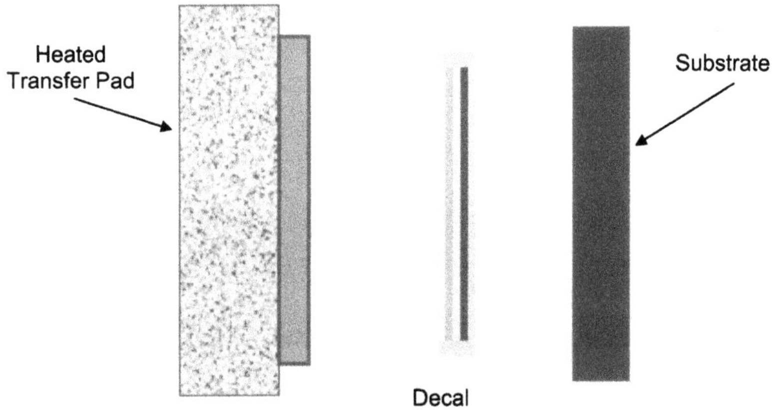

Figure 4. Transferring a Direct Heat Transfer Ceramic Decal

HOW DECALS CAN SOLVE DECORATING CHALLENGES

There are many different reasons why ceramic decals are used. Most often ceramic decals are selected as the method of decoration because they solve a particular decorating challenge. Compound curves are difficult to print by direct screening. A decal is flexible enough to follow the irregular shape of a compound curve.

Decals can offer exceptional registration. Printing onto paper, which is a flat, controlled and uniform surface, allows for precision registration. This is extremely important when printing halftone work used in fine detail and art reproduction work.

Inconsistent substrates can be impossible to register when printing multiple colors using direct screening. Decals assure excellent registration and minimal losses on difficult shapes.

Just-in-Time projects can benefit from using ceramic decals. Minimal setup times with decal make it efficient to run small batches. Finished goods inventory can be kept to a minimum by decorating common substrates with unique images only when there is demand.

Heat release decals can offer decal quality with fast production speeds. Other advantages include consistent image placement, the ability to fire immediately after decoration and quick setup.

Decals are often used when a full surface decoration is required. It is difficult to direct screen fully around a cylinder. Printing a very small overdraw of covercoat at the edge of a decal enables the decal to wrap together with a minimal seam.

Some substrates have recessed areas that are difficult to decorate. A decal can easily be placed in a recessed area or around a corner. The flexibility of a decal makes decorating irregular surfaces possible.

Some specialty uses of ceramic decals include having raised print. Raised enamel dots can be printed on a decal in lieu of decorating by hand. Other special effects utilizing raised print can include creating a texture to produce an imitation leather or fabric.

Decals should be stored in controlled conditions with a temperature of 72°F (22°C) and a relative humidity level of 50%. If the temperature is too high, the decals may stick together, ruining them. If the humidity level is too low, the solvents will escape from the decals, making them brittle when decorating. Decals should be stored on end, to keep pressure off of the printed surface. They should be stored in the original packaging to keep them clean and protected from the environment.

Waterslide decals should be applied in a clean workspace using fresh decorating water. Concentration of excess starch and gum will build up in the water if not changed frequently. The substrate should be free from any dirt or dust to minimize firing problems. Firm pressure should be applied to the top surface of the decal to evacuate all air and water from underneath the decal. Decorated pieces should be thoroughly dry before firing as any residual moisture will cause blowouts during the firing process.

All transfer pads and rollers should be cleaned regularly to prolong their life. Use the minimum amount of heat required to make a successful transfer. Excessive heat can damage the transfer pads and rollers. Heat transfer decals should not be stored in the same conditions as waterslide decals. Heat release decals generally do not have a slip sheet inserted between transfers, which make them much more susceptible to heat damage during transit and storage.

DIGITAL CERAMIC PRINTING

Ron Manwiller
Enduring Images

This paper reviews the overall ceramic ink jet printing technology and gives specific examples of state-of-the-art process equipment.

The dominant ceramic printing technology worldwide has been screen printing for over 50 years. In the next 3 to 5 years, digital ceramic printing will become a significant ceramic imaging manufacturing technology if not a complete replacement for most of what screen does today.

With digital ceramic printing, going from computer to print is simple–and getting easier. It is also a green process with all dry chemistry. There is no color separation, film, screens, or harsh organic solvents. There is essentially no waste stream. It has the highest resolution, 600 to 1200 dpi, of all commercial ceramic imaging technologies, and produces durable truly ceramic images using a state of the art printer platform with banner printing capability. Ceramic chemistry eliminates the substantial performance deficiencies of sublimation decorations. Fine particle size toners are offered in 6 formulations with custom formulations available.

Examples of the scope of the product line are:

- Photo porcelain "cameos" for the funeral industry
- Custom tile for bath, kitchen, swimming pool etc.
- High quality, durable customer recognition items with no minimum order
- Custom souvenirs in tourist locations
- Uniquely personal family heirloom gifts - baby, wedding, anniversary, birthday, graduation etc.
- Custom designed tableware

There are six steps in digital ceramic print. First, artwork is received as slides, negatives, digital photos, analog photos, computer graphics, camera ready art, paintings, finger paintings and many others. Second, it is digitized. Analog artwork can be scanned. Digital artwork can be received in essentially any format with .tif and .bmp preferred but .jpg is also acceptable. The minimum resolution for good quality is 300 dpi though 600 to 1200 dpi is possible. Third, it is modified in a program such as Adobe Photoshop to meet customer specifications. Fourth, it is printed. The digital file needs to be converted to CMYK format for digital printing then ceramic product images are formatted to fit the maximum number on a sheet. The image is next printed onto ceramic pigment-based water slide decal, which is applied to the porcelain item. The printed ceramic image is laminated to a transfer film which completes the production of a water slide decal. After, the image is cut to size and submersed in water which releases the image from the decal paper. It is positioned on the ceramic substrate and allowed to dry. Finally, the imaged ceramic is put into a kiln and fired to about 1600°F (870°C). During firing, the image is fired into the surface of the glaze for gloss and very high durability.

The resulting product delivers true, TCA tested ceramic performance – UV stability, scratch and abrasion resistance. If porcelain enameled metal is used, the item is exceptionally resistant to breakage in normal use.

A custom tile example is shown below. It is at the showroom of Glass Interiors Inc., and is a tile mural of a polar bear 48 inches wide by 96 inches high.

PRELIMINARY EVALUATIONS OF FUNCTIONAL COATINGS FOR IMPROVED CHEMICAL AND WATER RESISTANCE

Charles Baldwin, Lou Gazo
Ferro Corporation

ABSTRACT

Preliminary results of functional coatings to improve chemical resistance are reviewed.

INTRODUCTION

The purpose of this study was to investigate the feasibility of using functional coatings over porcelain enameled steel to improve chemical resistance.

Porcelain enamels for ferrous substrates are based on borosilicate fluxed with alkali and alkaline earths to allow the glass to fire out between 1400 and 1600°F (760 and 870°C). While the fluxes provide needed workability, there is also a trade-off in corrosion resistance.

Corrosion of glass exposed to an aqueous solution occurs in a two-stage process. First, ion exchange between alkali ions (lithium, sodium, or potassium) from the glass and hydrogen ions from the solution takes place. The rate of alkali extraction from the glass is parabolic, and the surface area increases because of the formation of micropores in the alkali-depleted surface layer. At acidic pH values, increased hydrogen cation concentration increases reaction rates. Second, after dealkalization, hydroxyl anions will directly attack the silicon oxide tetrahedra, causing total dissolution of the glass, which can also occur immediately in caustic solutions with pH values greater than 9.[1]

Two types of functional coatings were evaluated to reduce porcelain enamel corrosion, particularly in residential water heaters. The first was the application of a polytetrafluoroethylene (PTFE)-based non-stick coating to water heater enamel reformulated to act as a vitreous primer for the PTFE coating. The second was an alkali-free oxide sol-gel-derived thin film nanocoating.

ENAMEL/FLUOROPOLYMER COATING

Water heater enamels are high-refractory enamels specially formulated to withstand attack by a range of water chemistries at elevated temperatures in the range of 130 to 150°F (54 to 66°C). To compensate for holidays in the porcelain enamel coating, the water heater steel is also subjected to cathodic protection by a sacrificial aluminum or magnesium anode.[2]

The most meaningful accelerated assessment is by patch tank testing. The test uses a 30 gallon stainless steel water heater with 24 holes cut in the upper portion of the tank. Test panels are coated cut-outs from 30 gallon tanks that are mounted on gaskets at the holes to expose an area of 2 in^2 to softened water heated to 210 ± 10°F at a pressure of 30 to 40 psi. The tank discharges 10 gallons of water every three hours to use a total of 80 gal/day. Every three weeks, the weight loss is measured on the plates, and the test is ended after 24 weeks. For each test, a known industry standard with a statistically meaningful amount of historical data is ran.

Because of the high refractory load required in water enamels, firing temperatures of 1600°F (870°C) are typical. This firing temperature requires a heavier steel gauge to be used. Additionally, sandblasting is still common surface preparation for hot rolled steel tanks.

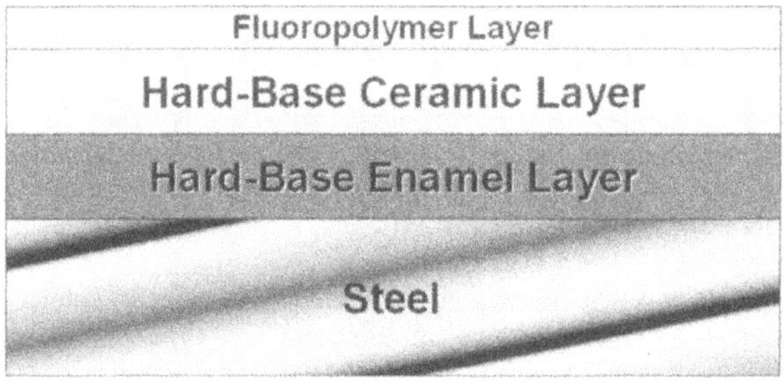

Figure 1. Steel hard-base schematic

The enamel/fluoropolymer coating system is shown schematically in Figure 1. The overall system allows a reduction in firing temperature versus typical water enamels as well as application to cleaned-only enameling-grade steel instead of blasted hot rolled steel.

First, a two-coat/one-fire enamel called steel hard-base was applied to patch tank test coupons and was to be used as a base coat for a highly chemically resistant fluoropolymer layer. Second, a high refractory ceramic coat was applied. Third, the hard-base was fired. The top surface was rough, permitting mechanical adhesion of a non-stick primer coating. Next, the multi-coat non-stick coating was applied and cured at 810°F. The steel hard-base is analogous to a previously described technology for improving the abrasion resistance of non-stick coatings on aluminum cookware.[3]

The hard-base serves as a dense, impermeable layer that prevents detachment of the organic coating by penetration of water to the base metal. This addressed the main failure mechanism of organic water heater coatings from detachment of large areas from cathodic disbanding induced by the sacrificial anode.[4]

Sample	Fire (°F/min)	Weight Loss (mg/in^2)
Industry Standard	1600/8	210.0
Hard-Base/Non-Stick	1450/6 and 810 F	25.6
Industry Sample	1560/6	224.6

Table 1. Water resistance of steel hard-base/non-stick system

Patch tank test results are shown in Table 1. The water resistance was improved by almost an order of magnitude at a lower firing temperature. Photographs of the tested plates in Figure 2 show the reduced visibility of the exposed area and less rust at the surface.

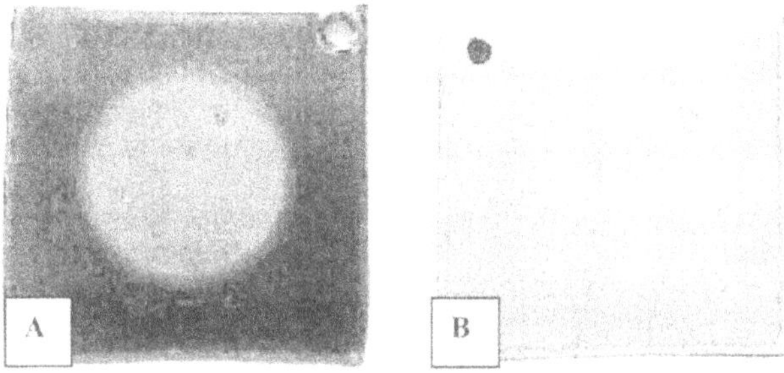

Figure 2. Appearance of tested (a) industry standard and (b) hard-base/non-stick patch tank samples

Figure 3 is optical micrographs of the surfaces of the industry standard sample and the PTFE sample before and after attack. The industry standard sample shows degradation of the porcelain at large bubbles first and also penetration of the water through the coating to the base steel.

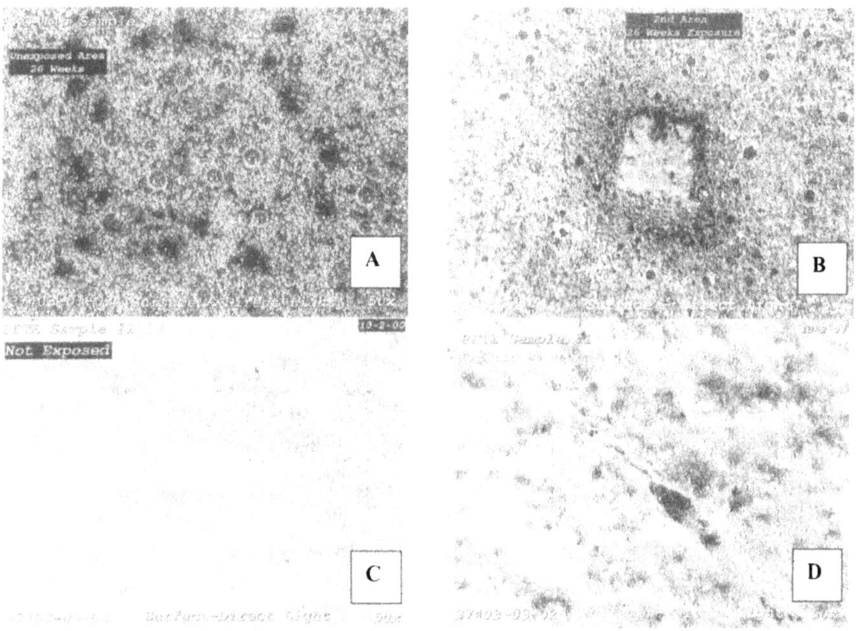

Figure 3. Optical micrographs of the industry standard for (a) an unexposed area and (b) an attacked area and (c) an unexposed area and (d) an attacked area of the PTFE sample.

Despite excellent corrosion resistance, PTFE coatings can contain residual porosity from processing and/or can be permeable to water. On a second set of test samples, the PTFE was purposely damaged by scribing an X with a razor across the face of the panel to assess the adhesion between the non-stick coating and the hard-base. Water did penetrate and undercut the adhesion layer. However, the hard-base would be left in place to protect the water heater steel.

Cross-sectional microscopy was done on samples that showed detachment. As shown in Figure 4, cracks were observed on polished cross-sections of the test samples. These may indicate adjustment of the relative thermal expansion of the ceramic and porcelain layers could be needed. Less permeable or less porous non-stick layers could also be investigated.

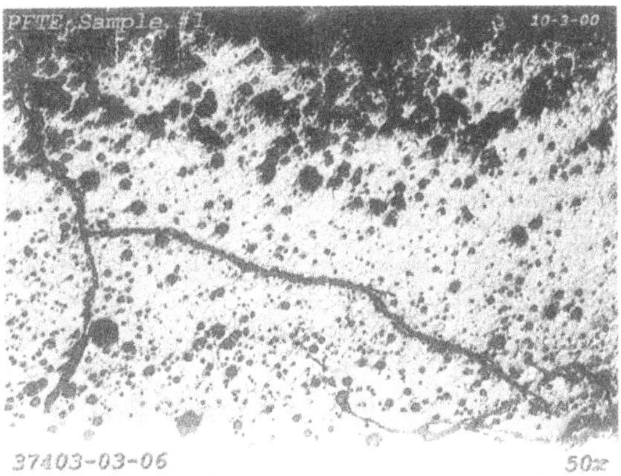

Figure 4. Optical micrograph of the cross-section of a tested PTFE patch tank panel

ENAMEL/NANOCOATING

An alternative approach was to coat the fired water heater steel with a sol-gel-derived nanocoating. A zirconia solution was added to a mixture of two alkyl silanes, allowed to hydrolyze, and then diluted in a carrier solvent. A single coat of the sol-gel was spray applied and cured at 750°F (400°C). Water resistance test results are shown in Table 2.

Sample	Fire (F/min)	Weight Loss (mg/in^2)
Industry Standard	1600/8	219.7
Enamel/Nanocoating Top-Coat	1600/8 and 750 F	110.5

Table 2. Water resistance of enamel/nanocoating system

To further evaluate the improvement in corrosion resistance, two-coat/one-fire electrostatically applied titanium white cover coat plates were coated with the nanocoating. Samples with and without the nanocoating were evaluated using boiling 6% citric acid (anhydrous) and 5% tetrasodium pyrophosphate aqueous solutions for 6 hours per ASTM C283-97. Three plates of each were tested. The average weight loss normalized to the exposed area are shown in Figure 5. The reduction in weight loss (and improvement in corrosion resistance) was very significant.

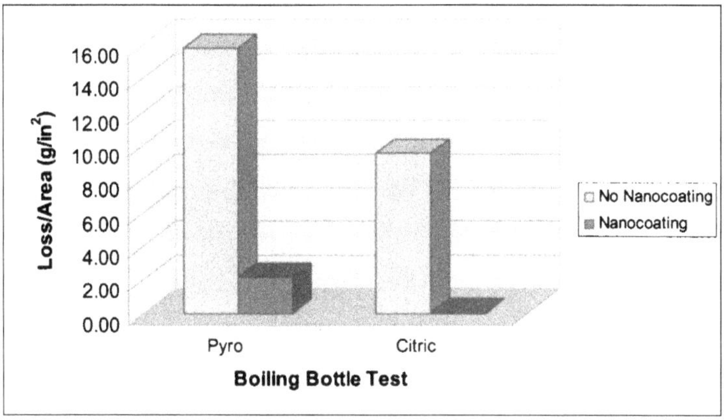

Figure 5. Boiling bottle results with nanocoating on white cover coat plates

SUMMARY

Based on preliminary results, the combination of a ceramic and fluoropolymer coating has promise to prolong the life of a water storage vessel. The structure of the coating contains a strongly bonded enamel layer lining the tank in the event of detachment of the fluoropolymer coating. Further evaluation of the adhesion of the fluoropolymer coating to the enamel hard-base in service, likely with electrochemical or cathodic delamination testing, is recommended. Using a less water permeable fluoropolymer such as PFA or FEP could also improve adhesion.[5]

Also based on preliminary results, alkali-free nanocoatings have promise to create thin, impermeable layers over porcelain enameled metal. The nano-scale particle size of the sol-gel process allowed firing at much lower temperatures than would be possible for an equivalent porcelain enamel composition. The silica/zirconia composition prevented alkali leaching in contact with corrosion aqueous solutions and significantly decreased weight loss of the samples.

ENDNOTES

[1] David E. Clark *et al.*, "Corrosion of Glass", Magazines for Industry: New York, NY, 1979.

[2] A. Saatchi *et al.,* "Investigation into Cathodic Protection Design for a Water Heater," *Materials Performance*, v. 42, n 3, pp. 20-24, 2003.

[3] William D. Faust, "Ceramic Substrate for Non-Stick Coatings," *Proceedings of the Porcelain Enamel Institute Technical Forum*, 63, 2001.

[4] L. Fedrizzi, "Study of the Corrosion Behavior of Phosphatized and Painted Industrial Water

Heaters," *Progress in Organic Coatings*, v 42, n 1-2, pp. 65-74, June 2001.

[5] Laurence W. McKeen, "Fluorinated Coatings and Finishes Handbook," William Andrews: Norwich, NY, 2006, pp. 321-353.

IMPROVING THE COVERING POWER OF TITANIUM WHITE COVER COAT BY CONTROLLING BUBBLE STRUCTURE

Koichi Tanaka
Tokan Material Technology Co., Ltd.

The color stability of a titanium white cover coat, relative to the enamel layer thickness, can be improved by controlling the bubble structure in the enamel.

Titanium white cover coat is widely used for electric appliances, cookware, architectural panels, etc. A variety of the shapes of the goods and the application methods tend to result in uneven enamel coating thickness, which sometimes causes defects in covering power. The thicker enamel coating can bring uniform color in general. There is a demand for a method to stabilize the color even with thinner or uneven titanium white coating thickness.

The goals of the study were to investigate whether bubble structure influences the color of titanium white cover coat and whether the color of titanium white cover coat can be stabilized by changing bubble structure.

Three different clays were evaluated. The composition of the clays is shown in Table 1.

	Clay A (wt%)	**Clay B (wt%)**	**Clay C (wt%)**
SiO_2	50.0	46.0	47.0
Al_2O_3	32.0	35.0	35.0
Na_2O	0.20	0.12	0.13
K_2O	2.00	0.76	0.76
Li_2O	0.04	0.03	0.04
CaO	0.30	0.20	0.22
MgO	0.20	0.30	0.22
lg-loss	10.00	12.00	12.61
H_2O	2.80	2.73	2.21

Table 1. Clay composition

The milling formula and specifications are shown in Table 2 for the first study of the effect of clay.

Material	A	B	C
Frit TMW-130	100	100	100
Clay A	6		
Clay B		6	
Clay C			6
Potassium Chloride	0.1	0.1	0.1
Water	50	50	50
Fineness	2-4[g/50cc/200mesh]	2-4[g/50cc/200mesh]	2-4[g/50cc/200mesh]
Specific Gravity	1.68-1.70 g/cc	1.68-1.70 g/cc	1.68-1.70 g/cc
Steel	JFE-CPE1D1	JFE-CPE1D1	JFE-CPE1D1
Spray Application	4, 5, 6, 7 [g/100*150mm]	4, 5, 6, 7 [g/100*150mm]	4, 5, 6, 7 [g/100*150mm]
Firing	1530°F (830°C) - 3min.	1530°F (830°C) - 3min.	1530°F (830°C) - 3min.

Table 2. Milling formulas for effect of clay

Figure 1 shows cross-sectional micrographs of the fired enamels.

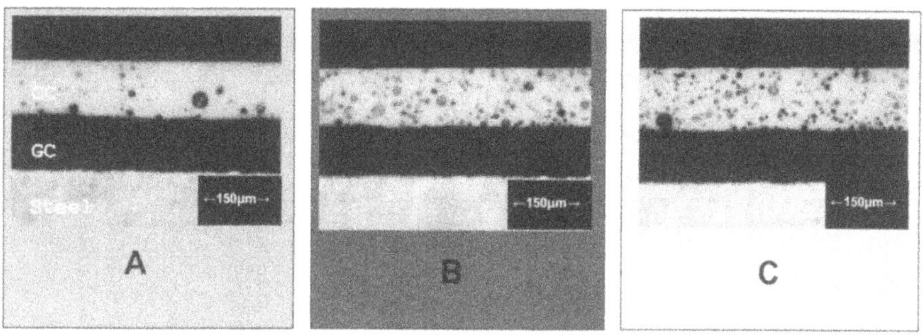

Figure 1. Cross-sectional micrographs with Clays A, B, and C

The size and number of bubbles, based on the micrographs, is shown in Figure 2.

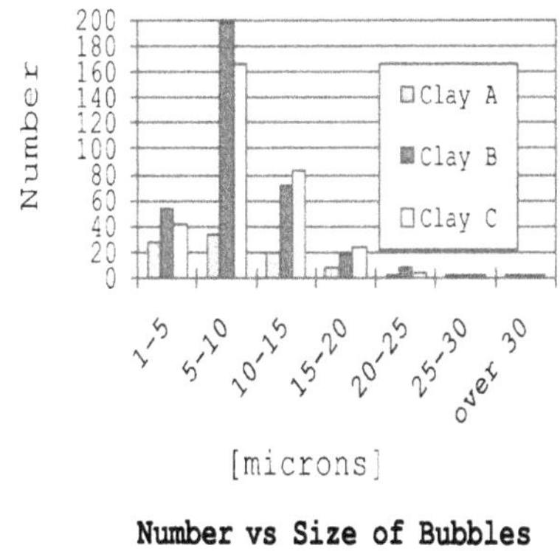

Figure 2. Number vs size of bubbles

The number of bubbles for Clay A was smaller. The bubble structures for Clay B and C are similar. This correlates with lower organic material content in Clay A shown in Table 1.

Figure 4 shows the reflectance (L value) versus coating thickness. Formula A showed significant color difference by thickness. Formulas B and C had consistent L values in all thicknesses.

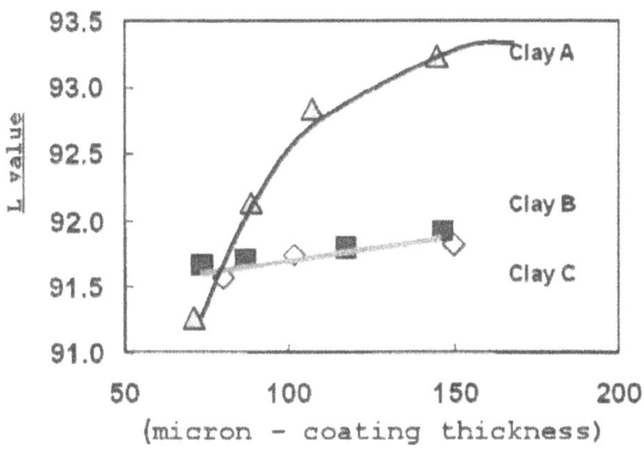

Figure 3. Reflectance versus coating thickness

Figure 4 shows coverage versus bubble structure. Thicker enamel with a coarser bubble structure had similar reflectance at different thickness values while thicker enamel with a finer bubble structure had different covering power versus thickness.

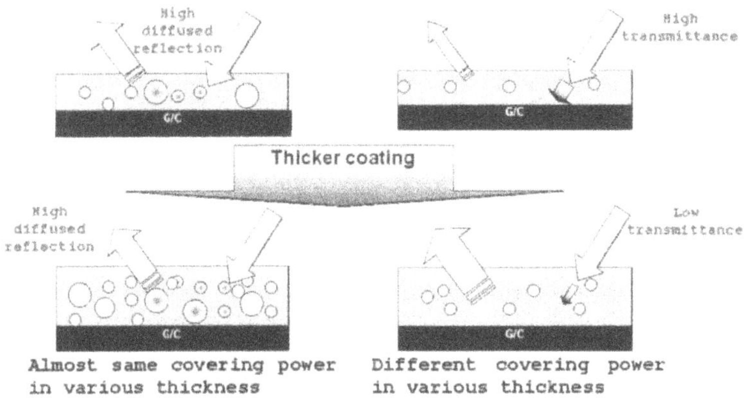

Figure 4. Coverage versus bubble structure

The milling formula and specifications are shown in for the second study of the effect of adding molybdenum trioxide (MoO_3).

Material	A	B	C
Frit TMW-130	100	100	100
Clay A	6		
Clay B		6	
Clay C			6
Potassium Chloride	0.1	0.1	0.1
Moly Trioxide (MoO_3)	0.0-0.5	0.0-0.5	0.0-0.5
Water	50	50	50
Fineness	2-4[g/50cc/200mesh]	2-4[g/50cc/200mesh]	2-4[g/50cc/200mesh]
Specific Gravity	1.68-1.70 g/cc	1.68-1.70 g/cc	1.68-1.70 g/cc
Steel	JFE-CPE1D1	JFE-CPE1D1	JFE-CPE1D1
Spray Application	4, 5, 6, 7 [g/100*150mm]	4, 5, 6, 7 [g/100*150mm]	4, 5, 6, 7 [g/100*150mm]
Firing	1530°F (830°C) - 3min.	1530°F (830°C) - 3min.	1530°F (830°C) - 3min.

Table 3. Milling formulas for effect of clay and molybdenum trioxide

Figure 5 shows cross-sectional micrographs of enamel with Clay A with the MoO_3.

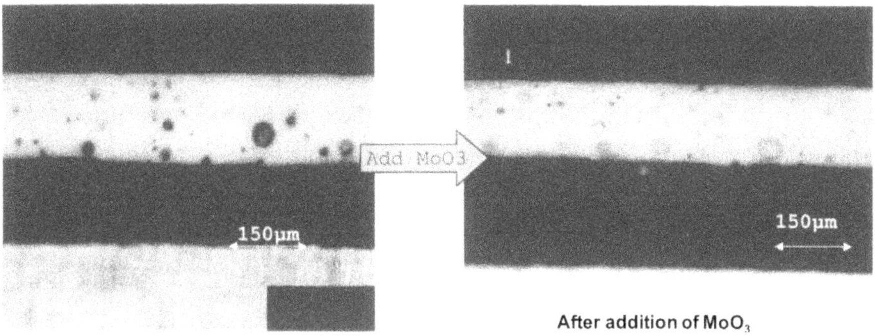

Figure 5. Micrographs of enamel with Clay A and added MoO_3

Figure 6 shows the results with Clay A and MoO_3. The added MoO_3 increased the bubble size and decreased the variation of L value with thickness.

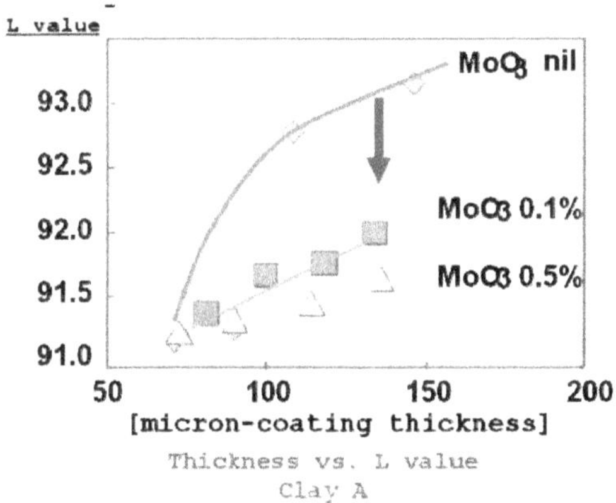

Size of bubble /µm	Clay A	Clay A +MoO3 (0.5%)
1-5	28	16
5-10	34	20
10-15	20	38
15-20	8	10
20-25	2	4
25-30	2	7
over 30	2	5
Total	86	100

Increase

Figure 6. Results with Clay A and MoO_3

Figure 7 shows cross-sectional micrographs of enamel with Clay B with the effect adding MoO_3. Bubble size decreased.

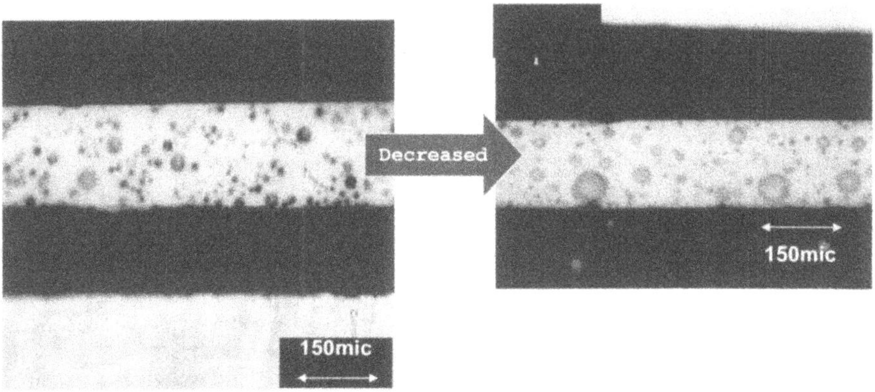

Figure 7. Micrographs of enamel with Clay B and added MoO_3

Figure 8 shows the results with Clay B and MoO_3. The added MoO_3 decreased the bubble size and increased the variation of L value with thickness.

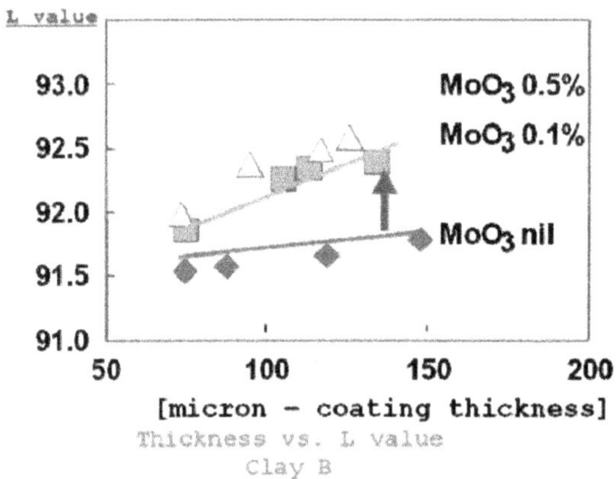

Size of bubble / µm	Clay B	Clay B +MoO_3 (0.5%)
1-5	53	0
5-10	199	11
10-15	71	43
15-20	17	49
20-25	6	37
25-30	1	11
over 30	1	15
Total	348	166

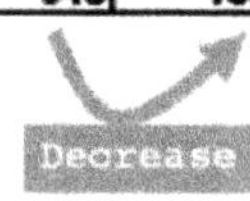

Figure 8. Results with Clay B and MoO_3

CONCLUSIONS

The bubble structure in enamel layer influences the stability of the titanium white color on different coating thickness. The bubbles in the enamel layer facilitated the light diffusion to stabilize the titanium white color on uneven coating surface with different thickness. The bubble structure can be controlled with additives. The color stability can be improved by controlling bubble structure in enamel.

CHEMICALLY RESISTANT ENAMEL

Eckhard Voß (Dipl.-Ing.)
Wendel GmbH

1. Introduction

Especially at low firing temperatures, it is difficult to achieve a good resistance against acids. When the firing temperature is approximately 700 °C, the possibilities are limited. The development presented in the following lecture provides a simple method to improve the resistance of low melting enamel. I would like to give an overview of various quartz modifications. The significant improvement of the chemical resistance of the enamels is confirmed by boiling tests. The application of the new batch compositions is demonstrated using a series of examples.

2. Improvement of the Chemical Resistance

To reach the optimum possible chemical resistance, enamels with appropriate composition are selected. Two possibilities to improve the chemical resistance and their related disadvantages are given in **Figure 1.**

1. **An increase of the resistant frit portion has the following disadvantages**
   1. Higher viscosity
   2. Bad wetting
   3. The enamel does not burn out thoroughly
   4. Higher price with equal viscosity
2. **An increase of the quantity of resistant mill batch additives has the following disadvantages**
   1. Matt surfaces
   2. Altered colour
   3. Poor adhesion

Figure 1

The chemical resistance of enamels is determined mainly by the content of quartz.

Enamels with an SiO_2 content higher than 60 % will become accordingly viscous and are difficult to process. The firing temperatures are above 850 °C and closed, flawless direct-on enamelling is hardly possible.

To achieve better processability, these enamels may have a higher alkali content. Lithium, in particular, enhances the chemical resistance and permits a lower firing temperature. As a side effect, however, the enamels become much more expensive.

To regulate the firing behaviour and the chemical resistance, it is a common practice to influence the properties by means of batch additives. The most frequently used additive is quartz in various grain sizes. Zirconium silicate and various other silicate compounds are also used.

These additives do not only make the enamels harder, they also change the brilliance and the colour.

With direct-on enamels, the additives also cause impairment of the adhesion.

3. Modifications of Quartz

The normal fireproof batch additive is quartz in the various grain sizes. Quartz is used to improve the chemical resistance and to control the firing behaviour.
If the enamels are fired at low temperatures, for example, 700 °C, there is only a limited choice of enamel frits with high chemical resistance. By addition of quartz, the enamels will become too hard quite easily.
In individual cases, solid colloidal silicon dioxide is added to the enamels. With majolica enamel, in particular, solid colloidal silicon dioxide is used to improve the chemical resistance. These fine quartz types also feature a certain suspension and set-up behaviour and change the rheologic properties. The maximum amount of additive is 4 %. Higher quantities cause surface flaws and make the enamel surface matt. The specific weight of these silicon dioxides is very low. The high volume of these powders also causes problems during mill filling.

In order to avoid the disadvantages mentioned above, the use of liquid colloidal silicon dioxide in enamels was investigated.

For comparison, tests using the 3 quartz modifications – fine quartz, solid colloidal silicon dioxide and liquid colloidal silicon dioxide - were made.
The mean grain size, the specific surface area (BET – Brunauer, Edward Teller) and the density are given in the table in **Figure 2**.

	Quartz W 500	Solid colloidal silicon dioxide	Liquid colloidal silicon dioxide
Mean grain size nm	13 000	16	8
Specific surface area m^2/g	1,36	110	330
Density g/cm^3	2,63	0,05	1,21

Figure 2

As shown in the table, liquid colloidal silicon dioxide has the smallest mean grain size. Compared to the finest quartz quality, the mean grain size of liquid colloidal silicon dioxide is by 1000 times lower.

The specific surface area is by 200 times larger compared to the one of fine quartz.

The density of the 3 quartz modifications varies widely.

Due to the large specific surface, in particular, a very high reactivity must be expected.
As shown in **Figure 3**, the diverging density has a considerable effect on the volume of the substances.
With an equal SiO_2 quantity, the volume of the solid nanoscale SiO_2 is very big. The inclusion of such a volume into the enamel batch will result in a change of the rheologic properties. The addition of quartz and liquid colloidal silicon dioxide does not cause any problems.

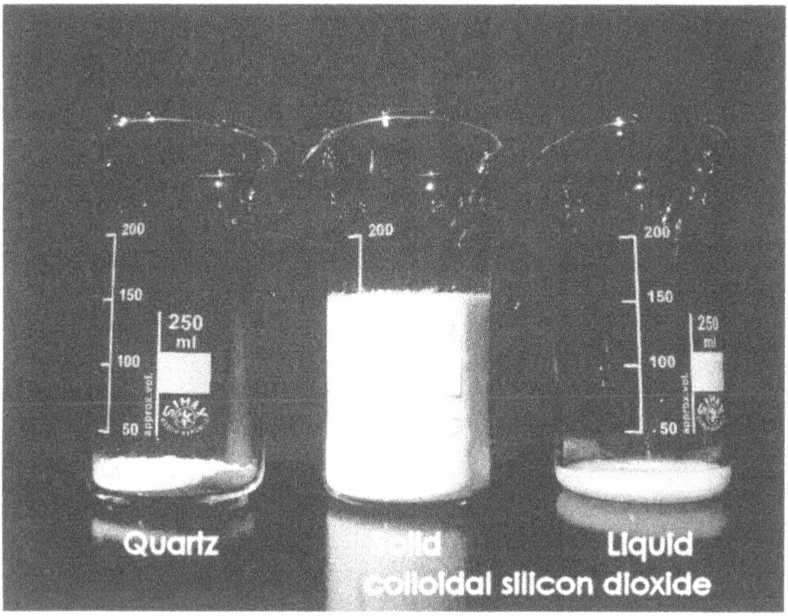

Figure 3

4. Enamel Batches with Liquid Colloidal Silicon Dioxide

Due to these properties of liquid colloidal silicon dioxide, the product can be used for enamel batches very easily. The advantages of liquid nanoscale silicon dioxide in the enamel batch are listed in **Figure 4.**

Liquid nanoscale silicon dioxide offers the following advantages:

- The portion can be calculated as a batch raw material
- High chemical resistance at low firing temperature
- Smooth surfaces
- Hard biscuit
- The surface remains wet during a longer time
- Easy improvement of the resistance against acids

Figure 4

Due to the high reactivity of the liquid nanoscale silicon dioxide, the quartz content of the enamel from the enamel recipe can be applied to the mill batch. The quartz from the raw material recipe can be used as a mill batch.
The quartz content and the water from the liquid colloidal silicon dioxide can be calculated and included into the batch.

Consequently, it is possible to use softer enamels and to fire them at lower temperatures. The chemical resistance corresponds to the one of considerably harder enamels.

Due to the low quantity of batch additives, very beautiful surfaces with high brilliance and smoothness are produced.

Portions of liquid colloidal silicon dioxide in the mill batches make the biscuit hard and touchproof. It is also possible to use the product in the ground enamel for the 2c/1f coating. This offers the advantage that the surfaces remain wet during a longer time.

The advantages become particularly obvious when considering the chemical resistance. In normal enamel batches, the mill quartz can be replaced by liquid nanoscale silicon dioxide. This is a simple method to improve the quality significantly.

5. Comparison of Chemical Resistances

Equal portions of 20 parts of quartz of various modifications were milled into a soft frit combination. For the investigations, fine quartz, nanoscale solid quartz and nanoscale liquid quartz were milled into a soft enamel with equal portions of SiO_2. The test plates were burned in at 700 °C during 6 minutes and tested in compliance with EN 14483-2-10. Testing according to EN 14483-2-10 is done using 6% citric acid boiling over 2.5 h.

Figure 5 very clearly reveals a significant improvement of the acid resistance due to the use of liquid colloidal silicon dioxide.

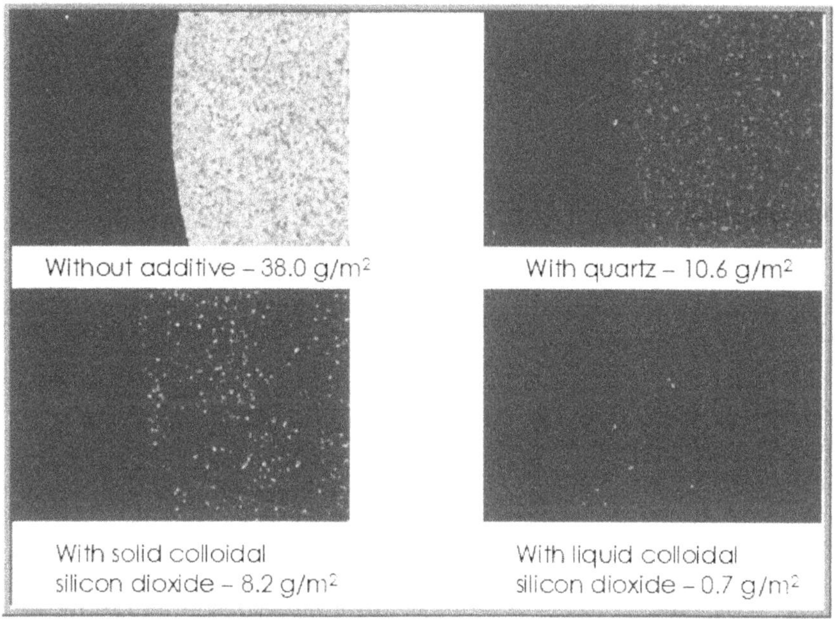

Figure 5

The enamel degradation is 38.0 g/m^2 without any additive at all, 10.6 g/m^2 with fine quartz, 8.2 g/m^2 with solid nanoscale quartz and 0.7 g/m^2 with liquid nanoscale quartz.
On the picture with liquid colloidal silicon dioxide, the homogeneous enamel structure is demonstrated very beautifully. On the picture relating to the solid colloidal silicon dioxide test, the surface is less homogeneous and reveals areas in which the citric acid has caused noticeable corrosion. Due to the large volume of solid nanoscale silicon dioxide, it is not possible to accomplish a homogeneous surface.
The liquid, nanoscale SiO_2 integrates itself into the enamel structure very homogeneously and improves the acid resistance considerably.

Figure 6 additionally confirms the improvement of the chemical resistance using the boiling test according to EN 14483-2-13. The boiling test according to EN 14483-2-13 is carried out using water over 48 h.

	Liquid phase g/m²*day	**Vapour phase** g/m²*day
Enamel without additive	4,8	26,0
Enamel with quartz	1,6	20,8
Enamel with solid colloidal silicon dioxide	1,7	18,8
Enamel with liquid colloidal silicon dioxide	0,2	21,9

Figure 6

After adding liquid colloidal silicon acid, the water resistance increases clearly. The picture shows very clearly that the liquid nanoscale silicon dioxide is included homogeneously into the enamel batch. The behaviour of fine quartz and solid nanoscale silicon dioxide relating to water resistance is equivalent. There are no differences in the degradation during the vapour phase due to the different quartz modifications.

6. Summary

The table in **Figure 7** gives a summary of the advantages of liquid nanoscale silicon dioxide.

- Easy improvement of chemical resistance
- Ideal for enamels with low firing temperature
- Quartz from the enamel recipe can be added using nanoscale SiO_2
- The mill water is added using the nanoscale SiO_2
- After application, the surface remains wet during a longer time
- Homogeneous enamel structure after firing

Figure 7

An improvement of the chemical resistance is achieved easily.

The usual quartz content of the mill batches can be replaced by liquid colloidal silicon dioxide simply by calculating the solid components. This does not influence the rheologic properties. A considerable improvement of the acid resistance, in particular, is obtained. The brilliance of the enamel is increased as well.

Very good suitability for enamels with low firing temperature.

At low firing temperature, fireproof batch additives tend to dissolve poorly in the enamel matrix. The melting temperature of the fireproof substances is far higher than the firing temperature of the enamels. For this reason, the enamel dissolution capacity is very low at low firing temperatures. Due to a particle size of 8 nm, however, liquid nanoscale silicon dioxide dissolves excellently in the enamel, also if the firing temperature and firing time are low.

Quartz from the enamel recipe can be added using liquid nanoscale silicon dioxide.

Hard chemically resistant enamels can be softened by reducing the quartz content in the enamel recipe. The reduced quartz content is added from the mill batch subsequently. Due to the nanoscale silicon dioxide particles, an enamel structure equivalent to the hard enamel type made from the melt is produced.

The mill water is added using the liquid nanoscale silicon dioxide.

As the mill batches all contain different quantities of water, the water from the liquid colloidal silicon dioxide can be taken into account as mill water when calculating. The rheologic properties are not changed due to the additive.

After application, the surface remains wet during a longer time.
Due to this property, the ground coat of enamel can be processed with liquid colloidal silicon dioxide using the 2coat/1fire method. The processing quality is improved considerably.

Homogeneous enamel structure after firing

Despite the short firing time and the low firing temperature, the nanoscale structure of the liquid silicone dioxide is integrated perfectly into the enamel matrix.

In summary, it can be said that liquid nanoscale silicon dioxide opens a large variety of possibilities for easy enhancement of the enamel properties.

THE WHO, WHAT, WHY, WHERE AND WHEN OF CAST IRON ENAMELING

Steve Pew
Porcelain Industries, Inc.

This paper will provide an overview of proper cast iron chemistries, processes and enamels for successful production parts.

INTRODUCTION

Porcelain Industries Inc. has been applying porcelain enamel to steel and cast iron for 52 years.

Elements of the cast iron enameling process are the substrate, enamel formulation, application, drying, and firing conditions.

For the substrate, the type and strength needs to be specified as well as the enamelability. Second, certain major constituents will play a significant role in the formulation as will whether the casting was made manually or automatically. The different pre-treatments before enameling will be reviewed as well as the configuration of commercial machines.

The appliance or cookware manufacturer specifies that the iron needs to have a yield strength of 25,000 psi and can be enamel coated. For the enameller, the iron must have good machining characteristics, be resistant to sag and creep during the firing process, and have a structure that is friendly to enameling with respect to surface defects. Surface defects can include voids, burn-off, chips, scaling, flaking, or bubbling.

Figure 1 shows the different types of ferrous alloys and cast iron. There are eight types of cast iron: (1) white, which is extremely hard crystalline structure resulting from solidification as cementite (Fe_3C), (2) gray, which is machineable, castable, enamelable with characteristics imparted from graphite structure type, (3) ductile, which is deformable without breaking used in pipe, (4) malleable, which is cast as white and heat treated to gain ductility used for pipe fittings, (5) chilled, which is white iron that has been produced by rapid solidification, (6) mottled that has intermediate characteristics to gray and chilled iron, (7) compared that is similar to gray but with coarser graphite, and (8) high alloy, which is specialized materials used for corrosion or temperature resistance.

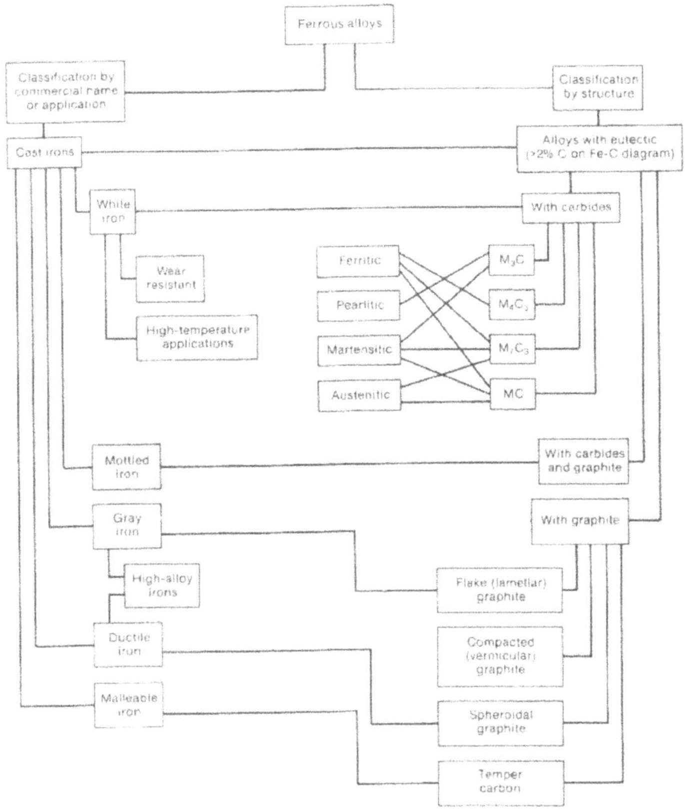

Figure 1. Types of ferrous alloys and cast irons

Figure 2 shows different types of cast iron versus carbon and silicon content. The range of compositions that can be enameled is highlighted in red.

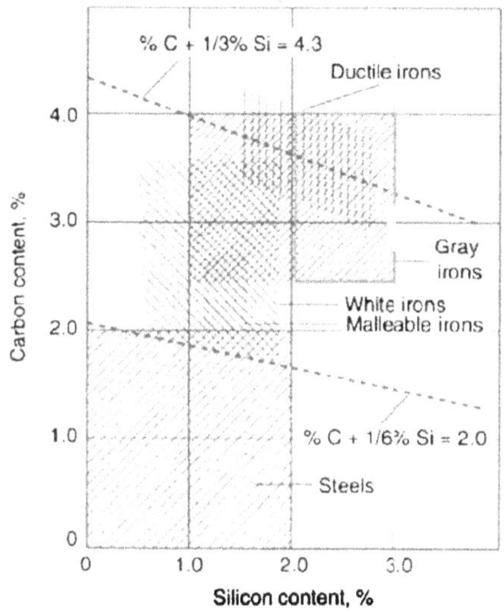

Figure 2. Overview of compositions of various cast iron types

Iron Composition

The chemical composition range to achieve enameling-grade gray cast iron is:

Total Carbon	3.25 - 3.60%
Graphitic Carbon	2.80 - 3.20%
Combined Carbon	0.22 - 0.52%
Silicon	2.25 – 3.00%
Manganese	0.45 – 0.65%
Phosphorous	0.60 – 0.95%
Sulfur	0.05 – 0.10%

The target microstructure is a well distributed fine to medium flake graphite in a matrix of ferrite or a ferrite – pearlite mix. This results in an enamelable casting with adequate strength and machineability.

Effects of Major Constituents

Carbon decreases the melting temperature of the iron to facilitate pouring. Note the eutectic point on the $Fe-Fe_3C-Si$ (2% Si) phase diagram. This allows melting at about 2100°F (1150°C). This is hundreds of degrees cooler than would otherwise be possible.

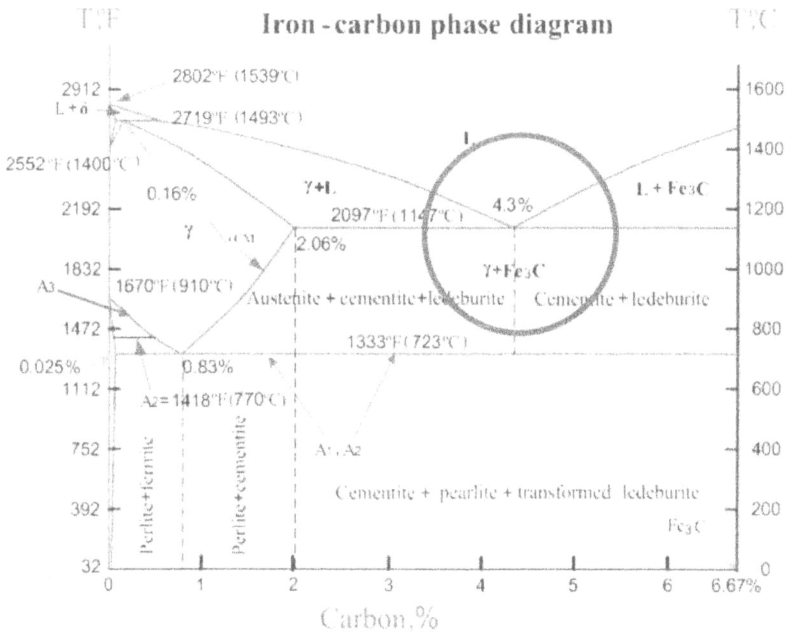

Figure 3. $Fe-Fe_3C-Si$ (2% Si) phase diagram[1]

Silicon

Due to the inter-reaction of carbon and silicon it is convenient to think of a combined factor called the carbon equivalent, CE:

Carbon Equivalent (CE) = %C + 1/3% Si

The CE describes how close a given formulation is to the eutectic composition (CE of 4.3). The CE lowers the eutectic point for a given carbon content, allows for casting thin sections of cookware or grate fingers, and affects the kinetics of carbide formation in favor of graphite.

Manganese

Manganese is a strong promoter of pearlite type matrix and inhibits the formation of iron sulfide (FeS) at grain boundaries by forming manganese sulfide (MnS) within the grains. For control of sulfur:

$$\%Mn \geq 1.17(\%S) + 0.15$$

Excess Mn combines with carbon to form stable manganese carbide (MnC), resulting in poor flow characteristics.

Phosphorous

Phosphorus promotes fluidity, which is helpful when filling long runs of thin sections. It may also be added to the CE calculation:

$$CE = \%C + (\%Si + \%P) / 3$$

Especially where formulations include appreciable amounts of P, such as cast iron formulations for pan supports.

Sulfur

Sulfur needs to be kept as low as possible! It counteracts silicon, acting as a carbide stabilizer. Sulfur can cause mis-runs due to the chilling effect, and it can cause pin-holing if not bound with Mn in the body of the grain.

Inoculation

Inoculation is a means of adding materials to the molten batch that act as nucleation sites for the formation of graphite during the cooling stage. This typically uses ferro-silicon often with additions of aluminum and calcium. The graphitization potential is increased during cooling and diminishes the tendency to form "chilled" iron. It also promotes a finer structure with a subsequent increase in strength.

Gray iron graphite types are classified into the following types:

- Type A - From inoculated iron with moderate cooling rate resulting in the best mechanical properties
- Type B - Limited nucleation sites available to promote precipitation of graphite, resulting in clusters of "rosettes"
- Type C – Formed from hypereutectic formulations with slow cooling rates
- Types D and E – Resulting from rapid cooling of hypoeutectic formulations

The distribution of graphite flake is shown in Figure 4.

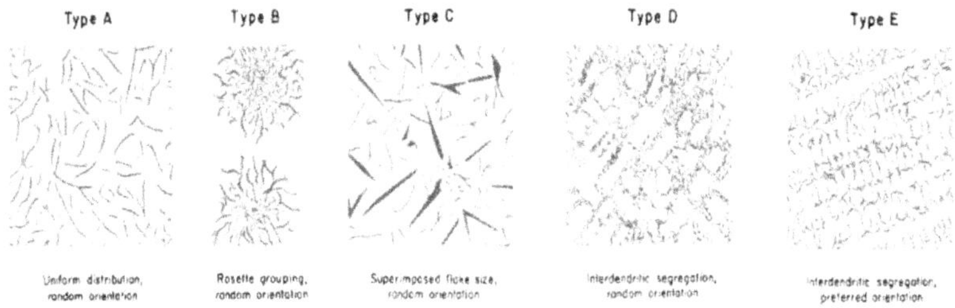

Figure 4. Types of gray iron graphite

Method of Casting

Considerations are whether to use permanent molds for sand casting or whether to use expendable molds as well as the sand composition and means of making the mold. The mold can be produced manually using match plate or cope and drag methods. Alternately, automatic molding can be done with vertical machines (Disa) or horizontal machines (Hunter).

The components of molding sand are the sand, clay, water, and coal powder. The sand can be fresh or recycled, as well as fine facing or more permeable body sand. The clay acts as a binder, and the water aids in the binding. The coal powder burns to create a gas layer between the iron and the sand. Percentages of constituents depend on complexity of the part, surface quality requirements.

Mold Making

For manual molding, Figure 5 shows the cope and drag method for a two piece pattern.

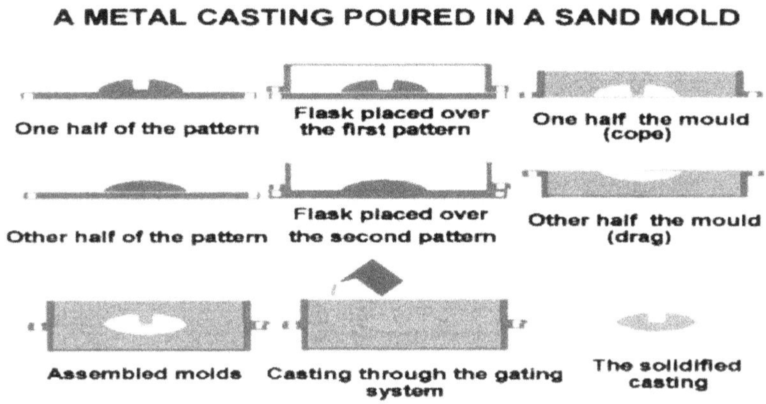

Figure 5. Cope and drag method for a two piece pattern

The match plate method is used for less challenging parts and has a one piece pattern with the top and bottom of the parts on either side of the pattern board.

Figure 6 shows the DISA Dansk Industri Syndikat A/S vertical mold machine. It is highly automated, very repeatable, but does not allow for facing sand. It uses short cooling cycles, and shake out is tough on delicate castings.

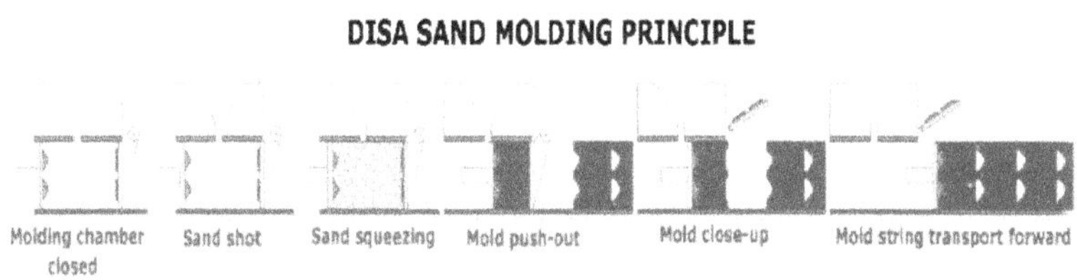

Figure 6. Disa vertical mold machine

Figure 7 shows a photo of a Hunter automated horizontal machine. It is a highly automated match plate type molding machine that forms molds in a horizontal plane.

Figure 7. Hunter (automated horizontal) molding machine

Methods of Surface Pretreatment

Surface pretreatment is needed to remove foundry sand and provide an appropriate surface for enameling. Different types are:

1) Mechanical abrasion with grit using a variety of machine configurations
    a) Sufficient speed to be economically viable
    b) Aggressive enough to yield suitable results
2) Abrasion from air blasted grit like aluminum oxide
    a) Short abrasive life due to fracturing of the particles
    b) Very hard on machine internally
    c) Yields a suitable surface for enamel coating
    d) Tends to be rather slow and, therefore, has unfavorable economics
3) Liquid/Abrasive mixtures pumped at high pressure
    a) Yields a surface suitable for enamel coating
    b) Tends to be very slow
    c) As yet unproven technology

The preferred method is mechanical abrasion using a centrifugal throwing wheel. There are many configurations of blasters that use the centrifugal wheel method of throwing media. For example, the Continuous Mesh Belt Blaster is for specialized application and is somewhat expensive. The Spinner Hanger is cost effective, widely used and relatively easy to maintain, and requires management of media impingement for best results. Tumble Blast is good for small pieces, inexpensive to buy and maintain, but tends to be rough on parts. A Table Blaster is good for low volumes, but uniform impingement is an issue.

Steel Shot is the first type of blast media. It is round spheres of steel tend to "peen" the surface, trapping contaminants below which eventually become enamel defects. Steel grit is the other and uses irregular shaped pieces of steel that abrade and remove material leaving a fresh surface to coat with enamel. The shot size and hardness are important. If it is too large, it will not get into the smaller crevices of the part. If it is too small, it will have insufficient kinetic energy to abrade the iron. If it is too soft, it rounds off too easily and behaves looks like shot. If it is too hard, it will do an aggressive job on the work piece as it consumes your machine. Figure 8 shows an SEM micrograph of a blasted cast iron surface.

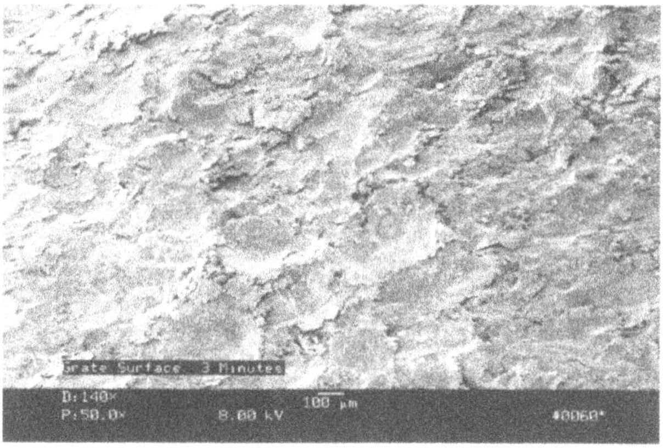

Figure 8. SEM micrograph of a cast iron surface after blasting with G-25 grit

This abraded surface yields characteristics suitable for enamel coating, such as sufficient surface roughness to provide mechanical adhesion of the coating to the iron and clean enough to preclude coating defects caused reactions of contaminants on the surface. The coating shown in Figure 9 had a smooth glossy appearance with good adhesion.

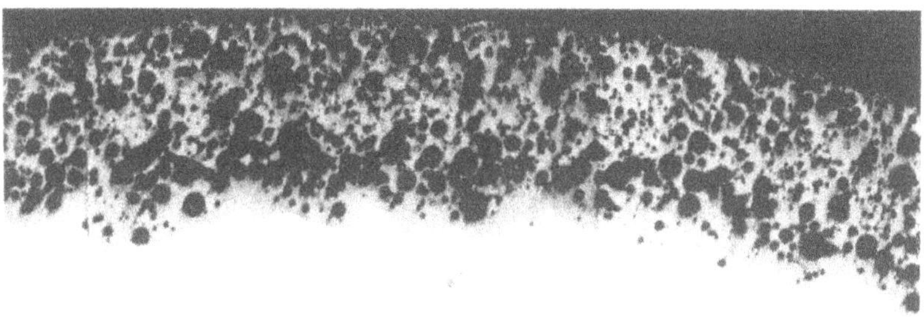

Figure 9. Micrograph of enameled cast iron

Enamel Formulation

The enamels must be formulated to achieve the particular attributes the end user requires such as acid resistance, thermal shock, craze resistance, color, etc. Gloss enamels are as fluid as practical at firing temperature. This allows outgassing to proceed to completion. Matt enamels are viscous enough at the firing temperature to hold down gas bubbles but fluid enough to achieve good mechanical adhesion.

Application

Any gun with wear resistant components (carbide) will work for spray application like Binks (95A with a 67VT nozzle, EIC (M1.8 fluid tip M4.7 air cap) or a Gremlin A28. Pump systems seem to work best. Enamel gravity is used that is suitable to avoid burn off and set to avoid drips. For dipping, use a good bucket and reclaim what you can!

Drying and Firing

The enamel needs to be dried slowly and completely, using lots of recirculation air and enough exhaust enough to keep the relative humidity in the dryer under control, and at as low of a temperature as possible.

Firing will depend greatly on the geometry of the ware being fired. It should be low and low, below 1400°F (760°C). Part soak times should exceed fifteen minutes, and the ware needs to be supported to control creep and sag. Use at least RA330 allow for tooling to control deterioration.

References

[1] "The Metallurgy of Cast Iron"
http://thdick.co.uk/index.php/process/the_metallurgy_of_cast_iron
(2 December 2009).

Author Index

www.ingramcontent.com/pod-product-compliance
Ingram Content Group UK Ltd.
Pitfield, Milton Keynes, MK11 3LW, UK
UKHW020817290926
15470UKWH00003B/1219

9 780470 408414